T0250055

Calculations in Chemical Kinetics for Undergraduates

Calculations in Chemical Kinetics for Undergraduates

Eli Usheunepa Yunana

CRC Press
Taylor & Francis Group
Boca Raton London New York

CRC Press is an imprint of the
Taylor & Francis Group, an **informa** business

First edition published 2022
by CRC Press
6000 Broken Sound Parkway NW, Suite 300, Boca Raton, FL 33487-2742

and by CRC Press
4 Park Square, Milton Park, Abingdon, Oxon, OX14 4RN

CRC Press is an imprint of Taylor & Francis Group, LLC

Library of Congress Cataloging-in-Publication Data

Names: Yunana, Eli Usheunepa, author.
Title: Calculations in chemical kinetics for undergraduates / Eli Usheunepa Yunana.
Description: Boca Raton : CRC Press, 2022. | Includes bibliographical references and index.
Identifiers: LCCN 2021059514 | ISBN 9781032228341 (hardback) | ISBN 9781032228204 (paperback) |ISBN 9781003274384 (ebook)
Subjects: LCSH: Chemical kinetics--Mathematics--Textbooks. | Chemical kinetics--Problems, exercises, etc. | Problem solving.
Classification: LCC QD502 .Y86 2022 | DDC 541/.394--dc23/eng20220314
LC record available at https://lccn.loc.gov/2021059514

ISBN: 9781032228341 (hbk)
ISBN: 9781032228204 (pbk)
ISBN: 9781003274384 (ebk)

DOI: 10.1201/9781003274384

Typeset in Times
by Deanta Global Publishing Services, Chennai, India

This book is dedicated first to God Almighty, the fountain of all wisdom. He has made it possible for what has remained in the realm of the mind to now be fully turned into tangible material. The time and resources invested into this book project over the years, He supplied because of His unwavering love. Secondly, I dedicate this book to the community of chemical scientists whose discoveries are immense foundations that set the pace for this project. No man can ever claim superior knowledge on a subject without following the already established findings made by others.

Contents

Preface

The adjective kinetic originates from the Greek word "kinetikos," which means "set in motion" or "move." Many chemical processes occurring around us, such as the rusting of iron, have been observed to take place over time. In the kinetic study of chemical processes, it is expedient to note these things – first, the rate at which the chemical reaction takes place, and second, the factors that affect the speed of the reaction. At this introductory level, the majority of study is devoted to how fast a reactant is consumed as opposed to how fast products are formed. It considers the chemical transport system which deals with the speed, velocity and rate of a chemical reaction.

Kinetics studies the rate of a chemical reaction, which means it is a time-based study. In every case, we will ask how fast something took place. In this introductory part of kinetics, we will study reactions that go from left to right. Later, it will get more complicated. The most typical thing in studying the kinetics of chemical processes is the change in concentration (molarity) over a given period. Note that concentration is the amount of a substance in each volume or quantity of solution; the unit is therefore mol per dm^3 (mol dm^{-3}). Here is a general form of chemical reactions: Reactants→Products.

This book is aimed at restoring passion for problem-solving and applied quantitative skills in undergraduate chemistry students. The fear of handling mathematically applied problems in physical chemistry topics has affected the foundation of the average chemistry undergraduate. This accounts for the poor background of many chemistry graduates who cannot connect fundamental theoretical chemistry to real experimental applications. Hence, the motivation for this book is to enhance the understanding of the average chemistry student.

This book contains fundamentals of chemical reactions, rate laws and methods of determining reaction order, theories and approximations in reaction kinetics and reaction mechanisms. All these contents define the scope of this book with emphasis on the undergraduate chemistry students as the target audience. However, the contents are rich and useful to even the graduate chemist who has a passion for applied problems in the physical chemistry of reaction kinetics.

Acknowledgment

I wish to acknowledge the immense contributions of some seasoned experts in academia whose relentless efforts in scrutinizing this work led to its successful compilation. To the esteemed mentors in my life, Prof. Eddy Nnabuk Okon, a seasoned computational and physical chemist both home and abroad and also Dr. Nicolas Andres Flores-Gonzalez, thank you for your enormous contributions in scrutinizing this work. My appreciation also goes to other academic mentors whose contributions I will not trivialize, Prof. Zakari Ladan, Prof. Inuwa M. Ibrahim and Mr. Nathaniel Atamas Bahago. And finally, to my beloved parents Mr. and Mrs. Yunana Alhassan, my siblings and my lovely fiancée Miss Melody Akuso, thank you all for your support and sacrifices.

Author

Eli Usheunepa Yunana is an inspiring teacher and mentor of high school and A-level chemistry students with seven years of experience as a tutor at City Academy, Maranatha College, and a volunteer tutor at the Department of Chemistry, Taraba State University (NYSC), all based in Nigeria. He is a solid-state materials chemist under the mentorship of Prof. Duncan H. Gregory, WestCHEM Chair in Inorganic Materials, School of Chemistry, University of Glasgow, United Kingdom.

Eli Usheunepa Yunana is a career-driven scientist and leadership mentor with a master's degree in advanced functional materials from the University of Glasgow. He is a member of ResearchGate, nanoHUB.org and an alumni student member of the Solid-State Group, also called the Gregory Group in the School of Chemistry, University of Glasgow. He has received outstanding recognition for excellence in his academic career as a chemist, ranging from the overall best undergraduate student, Kaduna State University (2018), to the award of a merit-based overseas scholarship by Kaduna State Government in Nigeria (2020).

He is currently the host of an online career and leadership mentoring platform (CAPLOM Services) with about 30 mentees and co-facilitators in leadership, career, financial intelligence, IT and programming courses. The platform has all its contents documented on different online media platforms. These reveal so much about the author as a vision-minded mentor with a great network of influence both in academic and leadership settings.

1 Reaction Rates, Molecularity and Rate Law

1.1 RATES OF CHEMICAL REACTIONS

This simply refers to the concentration of reactants used up to form products during a chemical reaction each time. Hence, the velocity of a chemical reaction is the rate at which the reactant concentration decreases or product concentration increases with time.

$$\text{rate} = \frac{\text{change in concentration}}{\text{time}} = \frac{\text{mol dm}^{-3}}{\text{s}} = \text{mol dm}^{-3}\,\text{s}^{-1}$$

Consider, the chemical reaction below.

$$aA + bB \rightarrow cC + dD$$

The rate of disappearance of a reactant = the rate of formation of a product. The reaction rate of the above chemical reaction is given below.

$$\text{Rate} = -\frac{\Delta[A]}{a\Delta t} = -\frac{\Delta[B]}{b\Delta t} = +\frac{\Delta[C]}{c\Delta t} = +\frac{\Delta[D]}{d\Delta t}$$

The negative sign indicates a decrease in the concentration of reactants while the positive sign indicates an increase in the concentration of products formed.

Δ represents the term "change" and a, b, c and d represent the stoichiometric coefficients for the reaction above.

Note: The rate of reaction must consider the number of moles of the individual reactants and products in the stoichiometric equation.

Example 1: Write the rate of reaction for the following chemical reactions.

(a) $CH_3COOH_{(aq)} + C_3H_7OH_{(aqs)} \rightarrow CH_3COOC_3H_{7(aqs)} + H_2O_{(aqs)}$

(b) $2HI_{(g)} \rightarrow H_{2(g)} + I_{2(g)}$

DOI: 10.1201/9781003274384-1

SOLUTION

Since Rate $= \dfrac{\text{change in concentration}}{\text{time}}$

(a) Rate $= -\dfrac{d[CH_3COOH]}{dt} = -\dfrac{d[C_3H_7OH]}{dt} = \dfrac{d[CH_3COOC_3H_7]}{dt} = \dfrac{d[H_2O]}{dt}$

(b) Rate $= -\dfrac{d[HI]}{2dt} = -\dfrac{d[H_2]}{dt} = \dfrac{d[I_2]}{dt}$

Example 2: Calculate the rate at which HI disappears in the reaction below when I_2 is being formed at the rate of 1.8×10^{-6} moles per litre per second.

$$2HI_{(g)} \rightarrow H_{2(g)} + I_{2(g)}$$

SOLUTION

Rate of disappearance of HI = rate of formation of I_2.

$$-\dfrac{d[HI]}{2dt} = \dfrac{d[I_2]}{dt} \qquad\qquad (1.1)$$

To obtain the rate of disappearance of 1 mole HI, we multiply the above rate expression by a factor of 2 on both sides.

$$-\dfrac{d[HI]}{dt} = \dfrac{2d[I_2]}{dt}$$

$$\text{Rate} = -2\left(1.8 \times 10^{-6}\right) \text{molL}^{-1}\text{s}^{-1}$$

$$= -3.6 \times 10^{-6}\ \text{molL}^{-1}\text{s}^{-1}$$

Example 3: Use the experimental data in the table below to answer these questions.

Concentration of phenolphthalein/ (mol dm³ s⁻¹)	5.0×10^{-3}	4.5×10^{-3}	4.0×10^{-3}	3.5×10^{-3}	3.0×10^{-3}
Time/(s)	0.00	10.5	22.3	35.7	51.1

Calculate the rate at which phenolphthalein reacts with the OH⁻ ions during each of the following periods.

(a) When the phenolphthalein concentration falls from 5.0×10^{-3} mol dm⁻³ to 4.5×10^{-3} mol dm⁻³.

(b) When the phenolphthalein concentration falls from 4.5×10^{-3} to 4.0×10^{-3} mol dm⁻³.

(c) When the phenolphthalein concentration falls from 4.0×10^{-3} to 3.0×10^{-3} mol dm⁻³.

SOLUTION

$$\text{Rate} = \frac{\text{change in phenolphthalein concentration}}{\text{change in time}}$$

$$= -\frac{\Delta[\text{phenolphthalien}]}{\Delta t}$$

(a)

$$= \left(\frac{0.0045 - 0.0050}{10.5 - 0.0}\right) \text{molL}^{-1}\text{s}^{-1}$$

$$= \left(-\frac{0.005}{10.5}\right) \text{molL}^{-1}\text{s}^{-1}$$

$$= 0.0000476 \,\text{molL}^{-1}\text{s}^{-1}$$

$$\text{Rate} = 4.76 \times 10^{-5} \,\text{molL}^{-1}\text{s}^{-1}$$

(b) $\text{Rate} = -\dfrac{\Delta[\text{phenolphthalein}]}{\Delta t}$

$$\text{Rate} = -\left(\frac{0.0040 - 0.0045}{22.3 - 10.5}\right) \text{molL}^{-1}\text{s}^{-1}$$

$$\text{Rate} = -\left(\frac{-0.0005}{11.8}\right) \text{molL}^{-1}\text{s}^{-1} = 0.0000424 \,\text{molL}^{-1}\text{s}^{-1}$$

$$\text{Rate} = 4.24 \times 10^{-5} \,\text{molL}^{-1}\text{s}^{-1}$$

(c) $\text{Rate} = -\dfrac{\Delta[\text{phenolphthalein}]}{\Delta t}$

$$\text{Rate} = -\left(\frac{0.0030 - 0.0040}{51.1 - 22.3}\right) \text{molL}^{-1}\text{s}^{-1}$$

$$\text{Rate} = -\left(\frac{-0.0010}{28.8}\right)\text{mol}L^{-1}s^{-1} = 0.0000347\,\text{mol}L^{-1}s^{-1}$$

$$\text{Rate} = 3.47 \times 10^{-5}\,\text{mol}L^{-1}s^{-1}$$

1.2   MOLECULARITY OF A REACTION

Some reactions occur in a single step. For example:

$$ClNO_{2(g)} + NO_{(g)} \rightarrow NO_{2(g)} + ClNO_{(g)}$$

Other reactions occur by a series of individual steps, for example, the mechanism of decomposition of N_2O_5.

Step I: $N_2O_5 \rightarrow NO_2 + NO_3$
Step II: $NO_2 + NO_3 \rightarrow NO_2 + NO + O_2$
Step III: $NO + NO_3 \rightarrow 2NO_2$

Hence, molecularity simply refers to the number of moles of reactants consumed in the rate-determining step of a reaction. It is usually a whole number. If the moles of reactant in the rate-determining step is one, it is called unimolecular. When it is two, it is bimolecular and in extreme cases, it could be three, hence it is said to be termolecular (trimolecular).

Note: The rate-determining stage of a reaction is the reaction that has the slowest rate. Hence, in determining molecularity, the mechanism and the slowest step of the reactions must be known.

Example: Determine the molecularity of the following reactions:

(a) $2H_{2(g)} + O_{2(g)} \rightarrow 2H_2O_{(g)}$

(b) $2NOCl_{(g)} \rightarrow 2NO_{(g)} + Cl_{2(g)}$

(c) $NO_{(g)} + O_{3(g)} \rightarrow NO_{2(g)} + O_{2(g)}$

SOLUTION

(a) 2 moles of H_2 + 1 mole of O_2 = 3 moles of reactants
    $\therefore$ Molecularity = 3 (termolecular).
(b) 2 moles of NOCl
    $\therefore$ Molecularity = 2 (bimolecular)
(c) 1 mole of NO + 1 mole of O_3 = 2 moles of reactants
    $\therefore$ Molecularity = 2 (bimolecular)

1.3   RATE CONSTANT AND RATE LAW

Rate law is the way the rate of a chemical reaction varies with reactant concentration. The rate equation is an experimentally determined equation that is consistent with the law of mass action in chemical reactions. This law states that the rate of a chemical reaction is directly proportional to the active masses of reactants.

Note: The active mass simply refers to the mathematical expression of concentration raised to exponential (power) of the number of moles corresponding to that of reactants. Remember, the stoichiometric coefficient is casually not equivalent to the exponential power represented as order. For instance, consider this reaction below.

$$mA + nB \rightarrow xC + yD$$

Applying the rate law based on the law of mass action.

$$\text{Rate} \propto [A]^m [B]^n$$

$$\text{Rate} = k[A]^m [B]^n$$

Where k is the rate constant, by implication, a measure of the rate of a reaction at a specified temperature and its unit depends on the order of a reaction.

Example 1: Write the rate law of the following reactions.

(a) $2HI_{(g)} \rightarrow H_{2(g)} + I_{2(g)}$

(b) $CH_3COO^-_{(aq)} + H_3O^+_{(aq)} \rightarrow CH_3COOH_{(aq)} + H_2O_{(l)}$

SOLUTION

(a) $\text{Rate} = k[\text{reactant}]^{\text{number of mole}}$

$$\text{Rate} = k[HI]^2$$

(b) $\text{Rate} = k[CH_3COO^-][H_3O^+]$

Example 2: Calculate the rate constant for the reaction between phenolphthalein and OH− ion if the instantaneous rate of reaction is 2.5×10^{-5} mol L^{-1} s^{-1} when the concentration of phenolphthalein is 2.5×10^{-3} mol dm^{-3}. Given the rate law to be (rate = k[phenolphthalein]).

SOLUTION

$$\text{Rate} = k[\text{phenolphthalein}]$$

$$\therefore k = \frac{\text{Rate}}{[\text{phenolphthalien}]}$$

Where

$$\text{Rate} = 2.5 \times 10^{-5} \, \text{molL}^{-1} \text{s}^{-1} \text{ and } [\text{phenolphthalein}]$$

$$= 2.5 \times 10^{-3} \, \text{M or mol L}^{-1}$$

$$k = \frac{2.5 \times 10^{-5} \, \text{molL}^{-1} \text{s}^{-1}}{2.5 \times 10^{-3} \, \text{molL}^{-1}}$$

$$k = 1 \times 10^{-5-(-3)} \, \text{s}^{-1}$$

$$k = 1 \times 10^{-2} \, \text{s}^{-1}$$

$$k = 0.01 \text{s}^{-1}$$

Example 3: What is the unit of the rate constant for a fifth-order reaction when the unit of concentration is mol dm^{-3} and that of time is in minutes?

SOLUTION

$$\text{Rate} = k[\text{reactant}]^5 \quad \left(\text{Rate law for a 5}^{\text{th}} \text{ order reaction}\right)$$

$$\frac{\text{concentration}}{\text{time}} = k[\text{reactant}]^{-5}$$

$$\frac{\text{mol dm}^{-3}}{\text{min}} = k\left[\text{mol dm}^{-3}\right]^5$$

$$\text{mol dm}^{-3} \, \text{min}^{-1} = k\left[\text{mol dm}^{-3}\right]^5$$

$$k = \frac{\text{mol dm}^{-3}\text{min}^{-1}}{\left[\text{mol dm}^{-3}\right]^{5}} = \frac{\text{min}^{-1}}{\text{mol}^{4}\left(\text{dm}^{-3}\right)^{4}}$$

$$k = \text{mol}^{-4}\left(\text{dm}^{3}\right)^{4}\text{min}^{-1}$$

EXERCISE 1

(1) For the reaction $H_{2(g)} + 2NO_{(g)} \rightarrow N_2O_{(g)} + H_2O_{(l)}$, if the rate at a particular instant at which nitric oxide (NO) disappears was found to be 1.0×10^{-2} mol $L^{-1} s^{-1}$, what is the simultaneous rate at which nitric oxide (N_2O) appears? $\left[\text{Ans} = 5.0 \times 10^{-3} \text{ mol } L^{-1} s^{-1}\right]$

(2) Write a balanced equation for the following reactions.

(i) Rate $= -\dfrac{d[N_2]}{dt} = -\dfrac{d[H_2]}{3dt} = \dfrac{d[NH_3]}{2dt}$

(ii) Rate $= -\dfrac{d[CH_4]}{dt} = -\dfrac{d[O_2]}{2dt} = \dfrac{d[H_2O]}{2dt} = \dfrac{d[CO_2]}{2dt}$.

(3) Given that the Rate $= k[A]^1[B]^2$ for a reaction, if the unit of concentration is in mol dm^{-3} and that of time is in minutes, what is the unit of the rate constant?

$$\left[\text{Ans} = \text{mol}^{-2}\left(\text{dm}^{3}\right)^{2}\text{min}^{-1}\right].$$

(4) Write the rate law of the following reaction.

$$N_{2(g)} + 3H_{2(g)} \rightarrow 2NH_{3(g)} \qquad \left[\text{Ans: Rate} = k[N_2][H_2]^3\right]$$

(5) The decomposition of nitrosyl chloride is as follows.

$$2NOCl_{(g)} \xrightarrow{\Delta} 2NO_{(g)} + Cl_{2(g)}$$

NOCL/(mol dm⁻³)	1.0×10^{-2}	7.1×10^{-3}	5.5×10^{-3}	4.5×10^{-3}	3.8×10^{-3}	3.3×10^{-3}
Time (min)	0	2	4	6	8	10

Using the above data, determine the average rate of disappearance of nitrosyl chloride.

(a) Between the second and fourth minute.

(b) Over the entire reaction.

$$\left[\text{Ans}:(a)\text{Rate} = 4.0 \times 10^{-3}\,\text{mol}\,\text{dm}^{-3}\,\text{min}^{-1} \quad (b)\text{Rate} = 3.4 \times 10^{-4}\,\text{mol}\,\text{dm}^{-3}\,\text{min}^{-1}\right].$$

(6) The rate of change of molar concentration of CH_3 radi-
cals in the reaction $2CH_{3\,(g)} \rightarrow CH_3{-}CH_3$ was reported as
$\dfrac{d[CH_3]}{dt} = -1.2\,\text{mol}\,\text{L}^{-1}\,\text{s}^{-1}$ under a particular condition. What is

(a) The rate of reaction?

(b) The rate of formation of $CH_3{-}CH_3$?

$$\left[\text{Ans}:(a)0.60\,\text{mol}\,\text{L}^{-1}\,\text{s}^{-1} \qquad (b)0.60\,\text{mol}\,\text{L}^{-1}\,\text{s}^{-1}\right].$$

(7) Given that the rate of formation of ammonia is 0.345 mol dm^{-3}s^{-1},
what is the rate of disappearance of H_2 as shown in the reaction
below?

$$N_{2(g)} + 3H_{2(g)} \rightarrow 2NH_{3(g)}$$

$$\left[\text{Ans}:-0.518\,\text{mol}\,\text{dm}^{-3}\,\text{s}^{-1}\right]$$

2 Determination of Reaction Order by Using Initial Rates

2.1 ORDER OF A REACTION

This is defined as the sum of the exponents of reactant concentration in the rate law, e.g. Rate $= k[A]^x[B]^y[B]^z$. The x, y and z are the reaction orders with respect to reactants A, B and C.

There are various orders of reactions which include:

(i) First-order reaction: When the rate of reaction is directly proportional to the concentration of reactant A as shown here, e.g. A → product; Rate=k [A].

(ii) Second-order reaction: When the rate is proportional to the square of the reactant concentration, e.g. 2A → product; Rate=k[A]² or A+B → product; Rate=k[A][B].

(iii) Third-order reactions: When the rate is proportional to the cube of the reactant concentration, e.g. $A + B + C \rightarrow$ product; Rate $= k[A][B][C]$ or $2A + B \rightarrow$ product; Rate $= k[A]^2[B]$ or $3A \rightarrow$ product; Rate $= k[A]^3$.

(iv) Zero-order reaction: When the rate remains constant regardless of the reactant concentration, e.g. $2HI \rightarrow H_2 + I_2$; Rate $= k[HI]^0$.

(v) Fractional-order reaction: When the exponent of the reactant concentration in a rate equation is a fraction, e.g. $2A \rightarrow$ Product; Rate $= k[A]^{1/2}$.

(vi) Pseudo-order reaction: When the order of a chemical reaction appears to be less than the true order due to experimental conditions, e.g.

$$CH_3COOCH_3 + H_2O \rightarrow CH_3COOH + CH_3OH \text{ where } [H_2O] \text{ is constant.}$$

$$Rate = k'[CH_3COOCH_3] \text{ where } k' = k[H_2O].$$

Note: In a real situation, the above reaction is a second-order reaction but due to the condition of the experiment, the reaction assumes a first-order

DOI: 10.1201/9781003274384-2

reaction as seen in the rate equation above. Hence, this reaction order is called pseudo-first-order. The word "pseudo" literally means false.

Example 1: Rate data were obtained for the following reaction:

$$A + 2B \rightarrow C + 2D$$

Experiment	Initial [A]/ (mol L⁻¹)	Initial [B]/ (mol L⁻¹)	Initial rate of formation of C/(mol L⁻¹ min⁻¹)
1	0.10	0.10	3.0×10^{-4}
2	0.30	0.30	9.0×10^{-4}
3	0.10	0.30	3.0×10^{-4}
4	0.20	0.40	6.0×10^{-4}

(a) Determine the order of the reactions and hence write the rate expression for this reaction.
(b) Calculate the rate of the reaction k.

SOLUTION

$$Rate = k[A]^x [B]^y$$

By comparing Experiments 1 and 3:

$$R_1 = k[A]_1^x [B]_1^y \qquad\qquad\qquad \text{eq (i)}$$

$$R_3 = k[A]_3^x [B]_3^y \qquad\qquad\qquad \text{eq (ii)}$$

Dividing equation (ii) by (i):

$$\frac{R_3}{R_1} = \frac{k[A]_3^x [B]_3^y}{k[A]_1^x [B]_1^y} \qquad\qquad \text{eq (iii)}$$

Note: Ensure the experiment considered is one in which only one reactant concentration varies while other reactant concentrations are kept constant.
    Substituting the data into equation (iii):

$$\frac{3.0\times10^{-4}}{3.0\times10^{-4}} = \frac{\cancel{(0.10)}^{x} (0.30)^y}{\cancel{(0.10)}^{x} (0.10)^y}$$

$$\frac{1}{1} = \left(\frac{0.30}{0.10}\right)^y$$

$$1 = 3^y$$

From indices $3^0 = 1$:

$$\therefore 3^0 = 3^y$$

$$y = 0$$

Reactant [B] has a zero-order.
   By comparing Experiments 2 and 3:

$$R_2 = k[A]_2^x [B]_2^y \qquad\qquad \text{eq (iv)}$$

$$R_3 = k[A]_3^x [B]_3^y \qquad\qquad \text{eq (v)}$$

Dividing equation (v) by (iv):

$$\frac{R_3}{R_2} = \frac{k[A]_2^x [B]_2^y}{k[A]_3^x [B]_3^y} \qquad\qquad \text{eq (vi)}$$

Substituting the data into equation (vi):

$$\frac{3.0 \times 10^{-4}}{9.0 \times 10^{-4}} = \frac{(0.30)^x (0.30)^y}{(0.10)^x (0.30)^y}$$

$$\frac{1}{3} = 3^x$$

From indices $\dfrac{1}{3} = 3^{-1}$:

$$3^x = 3^{-1}$$

$$x = -1$$

Reactant [A] has a negative first order.
   Hence, the rate expression is given as Rate $= k[A]^{-1} [B]^0$.

$$\text{Therefore, Rate} = \frac{k}{[A]}.$$

(a) From the rate expression:

$$k = \text{Rate} \times [A]$$

Using Experiment 4, where $R_4 = 6.0 \times 10^{-4}\,M\,min^{-1}$ $[A]_4 = 0.20M$:

$$k = 6.0 \times 10^{-4}\,M\,min^{-1} \times 0.20M$$

$$= 1.2 \times 10^{-4}\,M^2\,min^{-1}$$

Example 2

The initial rate of decomposition of acetaldehyde at 6,000 k, $CH_3CHO_{(g)} \rightarrow CH_{4(g)} + CO_{(g)}$ was measured at a series of concentrations with the following results.

$[CH_3CHO]/(mol\,dm^{-3})$	0.2	0.3	0.4	0.5	0.6	?
Rate $(mol\,dm^{-3}\,min^{-1})$	0.34	0.76	?	2.10	3.10	6.90

Using the above data,

(a) Determine the reaction order in the rate equation below.

$$\text{Rate} = k[CH_3CHO]$$

(b) Calculate the value of k.
(c) Calculate the rate of reaction when the concentration of acetaldehyde is 0.4 $mol\,dm^{-3}\,min^{-1}$
(d) Calculate the concentration of acetaldehyde when the rate of reaction is 6.9 $mol\,dm^{-3}\,min^{-1}$.

SOLUTION

(a) Rate $= k[CH_3CHO]^m$

By comparing Experiments 1 and 2 from the previous data:

$$R_1 = k[CH_3CHO]_1^m \qquad\qquad \text{eq (i)}$$

$$R_2 = k[CH_3CHO]_2^m \qquad\qquad \text{eq (ii)}$$

Dividing both equations:

$$\frac{R_2}{R_1} = \frac{k[CH_3CHO]_2^m}{k[CH_3CHO]_1^m}$$

$$\frac{R_2}{R_1} = \left(\frac{[CH_3CHO]_2}{[CH_3CHO]_1}\right)^m \qquad\qquad \text{eq (iii)}$$

Substituting the data into equation (iii):

$$\frac{0.76}{0.34} = \left(\frac{0.3}{0.2}\right)^m$$

$$2.24 = 1.5^m$$

Take the log of both sides.

$$\log 2.24 = \log 1.5^m$$

$$\frac{\log 2.24}{\log 1.5} = \frac{m\log 1.5}{\log 1.5}$$

$$m = \frac{\log 2.24}{\log 1.5} = 1.99$$

$$m \cong 2$$

Since the reaction is a second-order reaction, hence the rate law is written as

$$\text{Rate} = k[CH_3CHO]^2.$$

(b) To find k, we use equation (i):

$$R_1 = k[CH_3CHO]_1^m$$

Where $R_1 = 0.34, [CH_3CHO]_1 = 0.2, m = 2$.

$$k = \frac{R_1}{[CH_3CHO]_1^m}$$

$$k = \frac{0.34}{(0.2)^2} = \frac{0.34\,moldm^{-3}\,min^{-1}}{0.04(moldm^{-3})^2}$$

$$k = 8.5\,mol^{-1}\,dm^3\,min^{-1}$$

Hence, the rate equation is given by $R = 8.5[CH_3CHO]^2$.

(c) When $[CH_3CHO] = 0.4\,moldm^{-3}, \quad R = ?$

$$Rate = 8.5(0.4)^2$$

$$= 8.5 \times 0.16$$

$$= 1.36\,moldm^{-3}\,min^{-1}$$

(d) When the rate $R = 6.9\,moldm^{-3}\,min^{-1}, [CH_3CHO] = ?$

$$R = 8.5[CH_3CHO]^2$$

$$[CH_3CHO] = \sqrt{\frac{R}{8.5}}$$

$$[CH_3CHO] = \sqrt{\frac{6.9\,moldm^{-3}\,min^{-1}}{8.5\,mol^{-1}\,dm^3\,min^{-1}}}$$

$$[CH_3CHO] = \sqrt{0.8118(moldm^{-3})^2}$$

$$= 0.901moldm^{-3}$$

EXERCISE 2

(1) The following kinetic data were obtained for the reaction between hydrogen and chlorine gas at 600° C.

$$H_{2(g)} + Cl_{2(g)} \rightarrow 2HCl_{(g)}$$

Exp. no.	Initial [H_2] mol dm^{-3}	Initial [Cl_2] mol dm^{-3}	Initial rate/mol dm^{-3} s^{-1}
1	0.15	0.075	3.5×10^{-3}
2	0.30	0.15	1.4×10^{-2}
3	0.15	0.15	7.0×10^{-3}

(a) Deduce the order with respect to hydrogen and chlorine. Given the overall order for the reaction.

$$\left[\text{Ans} : [H_2] = 1^{st} \text{ order}; [Cl_2] = 1^{st} \text{ order}; \text{overall} = 2^{nd} \text{ order} \right]$$

(b) Write the rate law for the reaction.

$$\left[\text{Ans} : \text{Rate} = k[H_2][Cl_2] \right]$$

(c) Calculate the rate constant for the reaction. What is the unit of k?

$$\left[\text{Ans} : k = 3.11 \times 10^{-1}, \text{ unit of } k = mol^{-1} dm^3 s^{-1} \right]$$

(2) Consider the reaction: $SO_{2(g)} + O_{3(g)} \rightarrow SO_{3(g)} + O_{2(g)}$.
A rate study of the reaction was conducted at 298 K; the data that were obtained are shown in the table.

[SO_2]/mol L^{-1}	[O_3]/mol L^{-1}	Initial rate/mol L^{-1} s^{-1}
0.25	0.40	0.118
0.25	0.20	0.118
0.75	0.20	1.062

(a) What is the order with respect to SO_2 and O_3?

$$\left[\text{Ans} : [SO_2] = 2^{nd} \text{ order}; [O_3] = \text{zero order} \right]$$

(b) Write the rate law for this reaction:

$$\left[\text{Ans} : \text{Rate} = k[SO_2]^2 [O_3]^0 \right]$$

(c) Determine the value and units of the rate constant k.

$$\left[\text{Ans} : k = 1.89, \text{ unit of } k = mol L^{-1} s^{-1} \right]$$

(3) Given the following data, determine the rate law for the reaction.

$$2NO_{(g)} + Cl_{2(g)} \rightarrow 2NOCl_{(g)}$$

Exp. no.	[NO]/mol dm^{-3}	[Cl$_2$]/mol dm^{-3}	Rate/mol dm^{-3} s^{-1}
1	3.0×10^{-2}	1.0×10^{-2}	3.4×10^{-4}
2	1.5×10^{-2}	1.0×10^{-2}	8.5×10^{-5}
3	1.5×10^{-2}	4.0×10^{-2}	3.4×10^{-4}

Deduce the rate expression for the reaction:

$$\left[\text{Ans: Rate} = k[NO]^2[Cl_2] \right]$$

(4) Use the data below to determine:

Exp. no.	$[NH_4^+]/(mol\,dm^{-3})$	$[NO_2^-]/(mol\,dm^{-3})$	$Rate/(mol\,dm^{-3}\,s^{-1})$
1.	0.25	0.25	1.25×10^{-3}
2.	0.50	0.25	2.50×10^{-3}
3.	0.25	0.125	6.25×10^{-4}

$$NH_{4(aq)}^+ + NO_{2(aq)}^- \rightarrow N_{2(g)} + 2H_2O_{(aq)}$$

(a)   The rate law for the above reaction.

$$\left[\text{Ans: Rate} = [NH_4^+][NO_2^-] \right]$$

(b)   The rate constant k for the reaction.

$$\left[\text{Ans}: k = 2 \times 10^{-2}(mol\,dm^{-3})^{-1}\,s^{-1} \right]$$

(5) The initial rate data for the reaction $2N_2O_{5(g)} \rightarrow 4NO_{2(g)} + O_{2(g)}$ is shown in the following table. Determine the value of the rate constant for this reaction.

Exp. no.	[N$_2$O$_5$]/(mol dm^{-3})	Rate/(mol dm^{-3} s^{-2})
1	1.28×10^2	22.5
2	2.56×10^2	45.0

$$\left[\text{Ans}:k=1.76\times10^{-1}\,s^{-1}\right]$$

(6) (a) Determine the overall order of the reaction below for the following data.

$$H_{2(g)}+2ICl_{(g)}\rightarrow I_{2(g)}+2HCl_{(g)}$$

Exp. no.	P_{H_2} (Torr)	P_{ICl} (Torr)	Rate (Torr/s)
1	250	325	1.34
2	250	81	0.331
3	50	325	0.266

$$\left[\text{Ans}:\text{overall order}=2^{nd}\text{ order,}\quad\text{Rate}=kP_{H_2}P_{ICl}\right]$$

(b) What is the value and unit of k for the reaction above?

$$\left[\text{Ans}:k=1.65\times10^{-5}\text{ Torr}^{-1}\,s^{-1}\right]$$

3 Rate Law by Method of Integration and Half-Life

The method of using initial rates in rate law determination requires some data from the initial portion of multiple experiments. This method of integrated rate law uses the concentration–time relationship of an experiment. The integrated rate equation relates concentration and time for a given order of reaction by integrating the differential kinetic equation. This approach has been used to also calculate the half-life of a reactant. Hence, the half-life is the time taken for half of the reactant concentration to be converted into product and is represented as $t_{1/2}$.

Consider the general reaction,

$$aA \rightarrow products$$

We can write the rate of this reaction as follows.

$$Rate = \frac{-d[A]}{dt}\frac{1}{a} = k[A]^n$$

Where n = order of reaction with respect to reagent "A."

3.1 INTEGRATED RATE LAW FOR ZERO-ORDER REACTION

When $n = 0$, $\quad rate = \frac{-d[A]}{dt}\frac{1}{a} = k[A]^0$.

Note: "*a*" is a stoichiometric variable of the chemical reaction.

$$\frac{-d[A]}{dt}\frac{1}{a} = k$$

$$\frac{d[A]}{dt}\frac{1}{a} = -k$$

$$d[A] = -akdt$$

DOI: 10.1201/9781003274384-3

By integrating both sides:

$$\int_{[A]_0}^{[A]} d[A] = -ak \int_0^t dt$$

From integrals $\int dx = x \Rightarrow \int d[A] = [A]$ and $\int dt = t$:

$$[A] \Big|_{[A]_0}^{[A]} = -ak \Big|_0^t$$

$$[A] - [A]_0 = -akt - \left(-ak(0)\right)$$

$$[A] - [A]_0 = -akt.$$

$[A] = [A]_0 - akt$ (integrated rate law for zero order) here a=1; $[A] = [A]_0 - kt$

For the half-life of zero-order reactions, let half-life be written as $t_{1/2}$.

$$[A] = \frac{1}{2}[A]_0$$

By substituting these variables, we have:

$$\frac{1}{2}[A]_0 = [A]_0 - akt_{1/2}$$

$$akt_{1/2} = [A]_0 - \frac{1}{2}[A]_0$$

$$akt_{1/2} = \frac{1}{2}[A]_0$$

$$\boxed{t_{1/2} = \frac{[A]_0}{2ak}}$$ Where $a = 1; t_{1/2} = \dfrac{[A]_0}{2k}$

3.2 INTEGRATED RATE LAW FOR FIRST-ORDER REACTIONS

When $n = 1$,   rate $= -\dfrac{d[A]}{dt}\dfrac{1}{a} = k[A]^1$ where $a = 1$.

$$\frac{d[A]}{dt}\frac{1}{a} = -k[A]$$

$$\frac{d[A]}{dt} = -k[A]$$

$$\frac{d[A]}{[A]} = -kdt$$

Taking definite integrals of both sides:

$$\int_{[A]_0}^{[A]} \frac{d[A]}{[A]} = -k \int_0^t dt$$

From integration, $\int \frac{dx}{x} = \ln|x| \Rightarrow \int \frac{d[A]}{[A]} = \ln[A]$.

$$\ln[A] \Big|_{[A]_0}^{[A]} = -kt \Big|_0^t$$

$$\ln[A] - \ln[A]_0 = -k(t-0)$$

$$\ln \frac{[A]}{[A]_0} = -kt$$

From logarithms, $\ln e^N = x \Rightarrow N = e^x$.

Therefore; $\ln \frac{[A]}{[A]_0} = -kt$ can be written as

$$\frac{[A]}{[A]_0} = e^{-kt}$$

$$[A] = [A]_0 e^{-kt} \quad (\text{integrated rate law for first order})$$

or

$$\ln[A] = \ln[A]_0 - kt$$

For the half-life of first-order reaction, let half-life $t_{1/2}$ be measured in seconds or any other unit for time.

$$[A] = \frac{1}{2}[A]_0$$

By substituting these variables, we have:

$$\frac{1}{2}[A]_0 = [A]_0 e^{-kt_{1/2}}$$

Divide both sides by $[A]_0$.

$$\frac{1[A]_0}{2[A]_0} = \frac{[A]_0 e^{-kt_{1/2}}}{[A]_0}$$

$$\frac{1}{2} = e^{-kt_{1/2}}$$

Take the natural log of both sides.

$$\ln\left(\frac{1}{2}\right) = \ln e^{-kt_{1/2}}$$

From logarithm, $\ln_a N^x = x \ln_a N$.

$$\ln\left(\frac{1}{2}\right) = -kt_{1/2} \ln e$$

Also, $\ln e = 1$ from the logarithm:

$$\ln(2)^{-1} = -kt_{1/2}(1)$$

$$-1\ln 2 = -kt_{1/2}.$$

Multiply both sides by -1.

$$\ln 2 = kt_{1/2}$$

$$t_{1/2} = \frac{\ln 2}{k}$$

Since $\ln 2 = 0.693$,

$$\boxed{\therefore\ t_{\frac{1}{2}} = \frac{0.693}{k}}$$

3.3   INTEGRATED RATE LAW FOR SECOND-ORDER REACTIONS

When $n = 2$, $a = 1$, rate $= -\dfrac{d[A]}{dt} = k[A]^2$

$$\frac{d[A]}{dt} = -k[A]^2$$

$$\frac{d[A]}{[A]^2} = -kdt$$

Integral of both sides:

$$\int_{[A]_0}^{[A]} \frac{d[A]}{[A]^2} = -k\int_0^t dt.$$

$$\int_{[A]_0}^{[A]} \frac{1}{[A]^2} d[A] = -k\int_0^t dt$$

$$\int_{[A]_0}^{[A]} [A]^{-2} d[A] = -kt \Big|_0^t$$

$$\left[-\frac{1}{[A]} \Big|[A]_0^{[A]} \right] = -kt \Big|_0^t$$

$$\left[-\frac{1}{[A]} + \frac{1}{[A]_0} \right] = -k(t-0)$$

$$\left[-\frac{1}{[A]} + \frac{1}{[A]_0} \right] = -kt$$

Multiply both sides by -1.

Make [A] the subject of the formula.

$$\frac{1}{[A]} = \frac{kt}{1} + \frac{1}{[A]_0}$$

$$\frac{1}{[A]} = \frac{[A]_0 kt + 1}{[A]_0}$$

Take the inverse of both sides.

$$[A] = \frac{[A]_0}{[A]_0 kt + 1} \qquad \left(\text{Integrated rate law for } 2^{nd} \text{ order}\right)$$

For the half-life of second-order reactions, let half-life be $t_{1/2}$ in the integrated rate law:

$$[A] = \frac{[A]_0}{2}.$$

By substitution into the integrated rate law:

$$\frac{\cancel{[A]_0}}{2} = \frac{\cancel{[A]_0}}{[A]_0 kt_{1/2} + 1}$$

$$\frac{1}{2} = \frac{1}{[A]_0 kt_{1/2} + 1}$$

$$[A]_0 kt_{1/2} + 1 = 2$$

By making $t_{1/2}$ the subject of the formula we have,

$$[A]_0 kt_{1/2} = 2 - 1$$

$$[A]_0 kt_{1/2} = 1$$

Divide both sides by $[A]_0 k$.

$$\boxed{\therefore t_{1/2} = \frac{1}{[A]_0 k}}$$

3.4   EXAMPLES OF SOLVED PROBLEMS

Example 1: In the first-order reaction A→B, the initial concentration is 0.95 mol dm^{-3}. What will be the concentration after 30 seconds if its rate constant is 3.2 × 10^{-2} s^{-1}?

SOLUTION

Given k = 3.2 × 10^{-2} s^{-1}, $[A] = 0.95\,moldm^{-3}$, t = 30s, $[A]_f$ = ?

Using the integrated rate law equation for first-order reactions:

$$\ln[A] = \ln[A]_0 - kt$$

$$\ln[A] = \ln(0.95) - (3.2 \times 10^{-2}\,s^{-1} \times 30\,s)$$

$$\ln[A] = -0.051 - 0.96$$

$$\ln[A] = -1.011$$

$$[A] = e^{-1.011} = 0.36\,moldm^{-3}$$

Alternatively,

$$[A] = [A]_0\,e^{-kt}$$

$$= 0.95e^{-3.2 \times 10^{-2} \times 30}$$

$$= 0.95e^{-0.96}$$

$$= 0.95 \times 0.38$$

$$= 0.36\,moldm^{-3}$$

Example 2: A first-order reaction whose initial amount of reactant was 1.5 mol dm^{-3} decomposes into half of the initial amount. What is the time it took for the decomposition to occur given the rate constant is 1.2 × 10^{-3} s^{-1}?.

SOLUTION

Since the reaction is first-order and the final concentration was half the original concentration, the half-life of the first-order reaction is given as:

$$t_{1/2} = \frac{\ln 2}{k} = \frac{\ln 2}{1.2 \times 10^{-3} \text{s}^{-1}}$$

$$t_{1/2} = \frac{0.693}{1.2 \times 10^{-3}} = 577.67 \text{ s}$$

$$t_{1/2} = 577.67 \text{ s}$$

Example 3: The half-life for a second-order reaction is 48 s. What was the original concentration if its rate constant is $5.19 \times 10^{-2} \left(\text{mol dm}^{-3}\right)^{-1} \text{s}^{-1}$?

SOLUTION

The half-life of the second-order reaction is given below as:

$$t_{1/2} = \frac{1}{[A]_0 k}$$

By making $[A]_0$ the subject of the formula, $[A]_0 = \dfrac{1}{kt_{1/2}}$

$$[A]_0 = \frac{1}{5.19 \times 10^{-2} \times 48}$$

$$[A]_0 = \frac{1}{2.4912} = 0.40 \text{ mol dm}^{-3}$$

Example 4: The initial concentration of a reactant is 0.520 M and the rate constant k is 2.5×10^{-3} M⁻¹ s⁻¹. Calculate the final concentration of the reactant after 10 min.

SOLUTION

The unit of rate constant given is $M^{-1}\text{s}^{-1}$.
    By comparing the exponent of M with M^{1-n}, to determine the order of this reaction:

$$\cancel{M}^{-1} = \cancel{M}^{1-n}$$

$$-1 = 1 - n$$

$$n = 1 + 1$$

$$n = 2$$

Since n is 2, the reaction is second-order.

$$\frac{1}{[A]_f} = kt + \frac{1}{[A]_0} \quad \text{or} \quad [A] = \frac{[A]_0}{[A]_0 \, kt + 1}$$

$$[A]_0 = 0.52M, \quad t = 10 \, \text{min} = 10 \times 60 = 600 \, s, \quad k = 2.5 \times 10^{-3} \, M^{-1} s^{-1}$$

$$= 2.5 \times 10^{-3} \, M^{-1} s^{-1}$$

$$[A] = \frac{0.52M}{0.52M \times 2.5 \times 10^{-3} \, M^{-1} s^{-1} \times 600 \, s + 1}$$

$$[A] = \frac{0.52}{0.78 + 1}$$

$$[A] = \frac{0.52}{1.78} = 0.29M$$

$$[A] = 0.29M$$

Example 5: The graph of [A] versus time is a straight-line plot with a slope of –0.24 M/min. The initial concentration of reactant [A] is 0.820 M. Calculate the final concentration of reactant [A] after 180 s.

SOLUTION

A plot of [A] against time is a zero-order reaction. The integrated rate law for zero-order reactions is given as:

$$[A]_f = [A]_0 - kt$$

$$\text{slope} = -k = 0.24 \, M \, \text{min}^{-1}$$

$$[A] = 0.820M$$

$$t = 180s = 3 \, \text{min}.$$

$$[A]_f = ?$$

$$[A]_f = 0.820M - (0.24) M \cancel{\text{min}}^{-1} \times 3 \cancel{\text{min}}.$$

$$[A]_f = (0.820 - 0.72)M$$

$$[A]_f = 0.10M$$

EXERCISE 3

(1) The rate constant at 75° C for the decomposition of its first order is 0.636 per second. Determine the half-life of its reaction.

$$\left[\text{Ans}: t = 1.09\,s\right]$$

(2) A first-order reaction was 53% complete in 30 minutes at 20° C. What is the rate constant? $\left(\text{Given that } [A]_0 = 100\%\right)$

$$\left[\text{Ans}: k = 2.52 \times 10^{-2}\,\text{min}^{-1}\right]$$

(3) The decomposition that proceeds into two different types of gas under first order resulted in a certain amount of concentration from 0.397 M in 274° C. It was observed that its half-life is 19 s. Determine the amount after 100 s.

$$\left[\text{Ans}: 0.01M\right]$$

(4) The decomposition of NO_2 into NO and O_2 has a rate constant of 7.8×10^{-2} M s^{-1}. If the initial concentration of reactant is 0.56 M at 3.2 minutes, what is its final concentration?

$$\left[\text{Ans}: 0.06M\right]$$

(5) The half-life for a second-order reaction is 48 s. What was the original concentration if the rate constant is 5.19×10^{-2} M^{-1} s^{-1}?

$$\left[\text{Ans}: 0.40M\right]$$

(6) At a given temperature a second-order reaction has a rate constant of 2.5×10^{-3} M^{-1} s^{-1}. Calculate the time required (in seconds) for the reaction to be 63% complete.

$$\left[\text{Ans}: t = 681s\right]$$

(7) The reaction rate constant for a particular second-order reaction is 0.47 M^{-1} s^{-1}. If the initial concentration of the reaction is 0.53 M, how many seconds will it take for the concentration to decrease to 0.13 M?

$$\left[\text{Ans}: t = 12.34\,\text{s}\right]$$

4 Theoretical Models for Chemical Kinetic and Arrhenius Equations

There are fundamentally two models for studying the kinetics of chemical reactions. These are collision theory and transition state (activated complex) theory.

4.1 COLLISION THEORY

The collision theory is built upon three assumptions:

(i) The collision of reactant molecules is necessary for a reaction to occur.
(ii) The reactant molecules must have proper orientation for effective collision which results in chemical reactions.
(iii) A minimum amount of energy is required for bond breaking and bond formation in a reaction. This energy is referred to as activation energy E_a.

The chemical reactions below show the significance of molecular orientation for a reaction to proceed.

$$\underset{\text{Force}}{\underleftarrow{CH_3^{\delta+}-Br^{\delta-}}}+\underset{\text{Force}}{\underrightarrow{:\ddot{O}H^-}}\xrightarrow[\substack{\text{unfavorable}\\\text{reaction}}]{}CH_3-Br+OH^-$$

From the reaction above, the partially negative Br group will repel the negatively charged OH group, resulting in ineffective collision, hence the reaction is hindered.

$$\underset{\text{Force}}{\underrightarrow{:\ddot{O}H^-}}+\underset{\text{Force}}{\underleftarrow{CH_3^{\delta+}-Br^{\delta-}}}\xrightarrow[\substack{\text{favorable}\\\text{reaction}}]{}CH_3-OH+Br^-$$

From the reaction above, the partially positive CH_3 group will attract the negatively charged OH group adjacent to it. This attractive force will result in an effective collision and hence, the formation of CH_3-OH.

DOI: 10.1201/9781003274384-4

4.2   COLLISION RATES IN GAS MOLECULES

In the bimolecular collision of gases, there are factors that define chemical reactions, e.g. collision number (frequency) Z_{AB}, collision cross-section
(σ), average velocity of molecules $\bar{v}$, steric factor p and number of reactant
molecules n. In the derivation of collision frequency, all the above factors
are included. Considering the bimolecular reaction: $A + B \rightarrow$ products

$$-\frac{d\cap_A}{dt} = Z_{AB}e^{-E_a/RT} \qquad\qquad \text{eq (i)}$$

Where Z_{AB} = collision frequency and $e^{-E_a/RT}$ represents the Boltzmann's
factor (Figure 4.1).

$$d_{AB} = (r_A + r_B) = \text{collision diameter}$$

$$\sigma = \pi d_{AB}^2 \quad \text{or} \quad \pi(r_A + r_B)^2$$

$$Z_{AB} = \pi d_{AB}^2 \bar{V}_{AB} \cap_A \cap_B$$

$$= \sigma_{AB} \bar{V} \cap_B \cap_A$$

$$= \text{Area of cross section} \times \text{velocity} \times \text{total number of molecules}$$

where $\bar{V}_{AB} = \sqrt{\dfrac{8k_BT}{\pi\mu}}$ and μ = reduced mass $\quad \mu = \dfrac{m_A m_B}{m_A + m_B}$

$$Z_{AB} = \pi(r_A + r_B)^2 \left(\frac{8k_BT}{\pi\mu}\right)^{1/2} \cap_A \cap_B$$

$$k_B = \text{Boltzmann's constant}$$

$$k_B = 1.38 \times 10^{-23} \text{ JK}^{-1}$$

$$Z_{AB} = \pi(r_A + r_B)^2 \left(\frac{8k_BT}{\pi\mu}\right)^{1/2} \cap_A \cap_B \qquad\qquad \text{eq (ii)}$$

Substituting equation (ii) into equation (i):

$$-\frac{d\cap_A}{dt} = \pi(r_A + r_B)^2 \left(\frac{8k_BT}{\pi\mu}\right)^{1/2} \cap_A \cap_B e^{-E_a/RT} \qquad\qquad \text{eq (iii)}$$

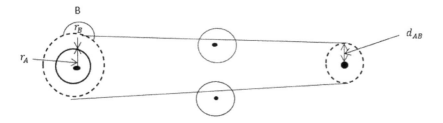

FIGURE 4.1 The reaction vessel and its geometry.

Since $[A] = \dfrac{\cap_A}{N_A}$ and $[B] = \dfrac{\cap_B}{N_B}$,

$$-\frac{d[A]}{dt} = k[A][B]$$

$$-\frac{d\cap_A}{N_A dt} = k\frac{\cap_A \cap_B}{(N_A)^2}$$

$$k = -\frac{d\cap_a}{dt} \times \frac{N_A}{\cap_B \cap_A}$$

$$k = \frac{N_A}{\cap_B \cap_A} \times -\frac{d\cap_a}{dt} \qquad \text{eq (iv)}$$

Substituting equation (iii) into equation (iv),

$$k = -\frac{N_A}{\cap_B \cap_A} \times \pi(r_A + r_B)^2 \left(\frac{8k_B T}{\pi\mu}\right)^{1/2} \cap_A \cap_B e^{-E_a/RT}$$

$$k = N_A \pi(r_A + r_B)^2 \left(\frac{8k_B T}{\pi\mu}\right)^{1/2} e^{-E_a/RT} \qquad \text{eq (v)}$$

The above equation can be condensed into the form

$$k = Z_{AB} e^{-E_a/RT}$$

When compared with the Arrhenius equation,

$$k = Ae^{-E_a/RT}$$

$$A = Z_{AB} \text{ but } A \neq Z_{AB} \text{ from experiment}$$

$A \neq Z_{AB}$ due to molecular arrangement in reactant molecules which creates steric hindrance during chemical interactions.

This implies that $\dfrac{A}{Z_{AB}} = p$; $(0 < p < 1)$

Where p = Steric factor

$f = e^{-Ea/RT}$ (molecular fraction colliding with activation energy).

Finally, the collision frequency for a bimolecular reaction of dissimilar species, Z_{AB}, from equation (v) is given by

$$Z_{AB} = \pi \left(r_A + r_B \right)^2 \left(\frac{8k_BT}{\pi\mu_{AB}} \right)^{1/2} N_A \qquad \text{eq (vi)}$$

$$Z_{AA} = 4\pi r_A^2 \left(\frac{8k_BT}{\pi\mu_{AA}} \right)^{1/2} N_A \qquad \text{eq (vii)}$$

The above equation (vii) is used to calculate the collision frequency of the bimolecular reaction of similar species for a reaction such as $A + A \rightarrow$ Product

$$k = Zpf \qquad \text{eq (viii)}$$

Where p = steric factor, Z = collision frequency.

4.3   TRANSITION STATE THEORY

In the transition state theory, an activated complex is proposed to exist between reactants and products in chemical reactions. This implies that reactants do not directly become products until an amount of energy is acquired to reach the transition state or activated complex. The activated complex is a short-lived specie because it is highly reactive (i.e. may break apart to form products or revert to reactants). The minimum energy required to get a reaction started is known as the activation energy E_a. Even if reactant molecules have the right orientation when colliding and do not have enough energy to reach the transition state, then the reaction will not proceed.

$$\Delta H = E_{a(forward)} - E_{a(reverse)}$$

From the above equation, the energy difference of the activated complex and reactants energy is the forward activation energy $[E_{a\ (forward)}]$. The energy difference of the activated complex and products energy is the reverse activation energy [$E_{a\ (reverse)}$]. The most important parameter for studying the energy changes in chemical reactions is enthalpy (ΔH).

4.4   ARRHENIUS EQUATION

The Arrhenius equation shows the temperature dependence of reaction rates. It is mathematically derived from $k = Zpe^{-E_a/RT}$ where Z=collision frequency and p=steric factor since $A = Zp$ $k = Ae^{-E_a/RT}$ where A=pre-exponential (frequency) factor.

By taking the logarithm of both sides,

$$\ln k = \ln A + \ln e^{-E_a/RT}$$

$$\ln k = \ln A - \frac{E_a}{RT} \ln e$$

Since ln e = 1

$$\ln k = \ln A - \frac{E_a}{RT}$$

$$\ln k = -\frac{E_a}{RT} + \ln A \quad \text{Arrhenius equation,} \qquad \text{eq(i)}$$

The Arrhenius equation can relate rate constants at two different temperatures.

At T_1 and k_1:

$$\ln k_1 = -\frac{E_a}{RT_1} + \ln A \qquad \text{eq (ii)}$$

At T_2 and k_2:

$$\ln k_2 = -\frac{E_a}{RT_2} + \ln A \qquad \text{eq (iii)}$$

Subtracting equation (ii) from equation (iii):

$$\ln k_2 - \ln k_1 = \left[-\frac{E_a}{RT_2} - \left(\frac{E_a}{RT_1} \right) \right] + \left(\ln A - \ln A \right)$$

$$\ln \left(\frac{k_2}{k_1} \right) = -\frac{E_a}{R} \left[\frac{1}{T_2} - \frac{1}{T_1} \right]$$

Taking the exponential of the right-hand side:

$$\frac{k_2}{k_1} = e^{-\frac{E_a}{RT}} \left[\frac{1}{T_2} - \frac{1}{T_1} \right]$$ eq (iv)

$$\boxed{k_2 = k_1\ e^{-\frac{E_a}{R}} \left[\frac{1}{T_2} - \frac{1}{T_1} \right]}$$

4.5   EXAMPLES OF SOLVED PROBLEMS

Example 1: A certain reaction has an activation energy of 50 KJ mol^{-1}. The rate constant at 25° C is 3.9 × 10^{-3} min^{-1}. What is the value of the rate constants at 75° C?

SOLUTION

Given that $E_a = 50$ KJ mol^{-1}, $R = 8.314$ J mol^{-1}
   $T_1 = 25°$ C $= 298$ K
   $k_1 = 3.9 \times 10^{-3}$ min^{-1}
   $T_2 = 75°$ C $= 348$ K
   $k_2 = ?$
Using the formula:

$$k_2 = k_1 e^{-\frac{E_a}{R}} \left[\frac{1}{T_2} - \frac{1}{T_1} \right]$$

$$k_2 = 0.0039 e^{-\frac{50,000}{8.314}} \left[\frac{1}{8.314} - \frac{1}{298} \right]$$

$$k_2 = 0.0039 e^{-6013.95} (0.00287 - 0.00336)$$

$$k_2 = 0.0039 e^{-6013.95} (-0.00049)$$

$$k_2 = 0.0039 e^{2.947}$$

$$k_2 = 0.0039 \times 19,049$$

$$k_2 = 7.4 \times 10^{-2} \text{ min}^{-1}$$

Example 2: The rate constants at 300 K and 500 K are 0.018 min^{-1} and 0.25 min^{-1}. Calculate the activation energy in KJmol^{-1}.

SOLUTION

$k_1 = 0.018$ min^{-1}, $T_1 = 300$ K
$k_2 = 0.25$ min^{-1}, $T_2 = 500$ K
Using the equation:

$$\ln\left(\frac{k_2}{k_1}\right) = \frac{-E_a}{R}\left[\frac{1}{T_2} - \frac{1}{T_1}\right]$$

By making E_a the subject of the formula:

$$-E_a\left[\frac{1}{T_2} - \frac{1}{T_1}\right] = R\ln\left(\frac{k_2}{k_1}\right)$$

$$-E_a = R\ln\frac{\left(\frac{k_2}{k_1}\right)}{\left(\frac{1}{T_2} - \frac{1}{T_1}\right)}$$

$$E_a = \frac{-R\ln\left(\frac{k_2}{k_1}\right)}{\left(\frac{1}{T_2} - \frac{1}{T_1}\right)}$$

$$E_a = \frac{-8.314\ln\left(\frac{0.25}{0.018}\right)}{\left(\frac{1}{500} - \frac{1}{300}\right)} = \frac{-8.314 \times 2.631}{(0.002 - 0.0033)}$$

$$E_a = \frac{-21.87}{-0.0013}$$

$$E_a = 16823.1 \text{Jmol}^{-1}$$

$$E_a = 16.8 \times 10^3 \text{ Jmol}^{-1}$$

$$E_a = 16.8 \text{KJmol}^{-1}$$

Example 3: The rate of a reaction triples when the temperature is increased from 25° C to 35° C. Calculate the activation energy for this reaction.

SOLUTION

Let the initial rate constant $k_1 = k$.
Therefore, the final rate constant $k_2 = 3k$.

$$T_1 = 25°C = 25 + 273 = 298K$$

$$T_2 = 35°C = 35 + 273 = 308K$$

$$E_a = \frac{-R\ln(k_2/k_1)}{\left(\dfrac{1}{T_2} - \dfrac{1}{T_1}\right)}$$

$$E_a = \frac{-8.3145\ln(3k/k)}{\left(\dfrac{1}{308} - \dfrac{1}{298}\right)}$$

$$E_a = \frac{-8.3145\ln 3}{0.00325 - 0.00336} = \frac{-8.3145\ln 3}{-0.00011}$$

$$E_a = 83040.1 Jmol^{-1}$$

$$E_a = 83.04 \times 10^3 \, Jmol^{-1}$$

$$E_a = 83.04 KJmol^{-1}$$

Example 4:

(a) (i) Calculate the collision rate for the decomposition of hydrogen iodide as shown in the equation below.

$$2HI_{(g)} \rightarrow H_{2(g)} + I_{2(g)}$$

(a) (ii) Calculate the fraction of molecules that are activated when the activated energy for the reaction at 600 K is 184.4 J mol^{-1}, the collision diameter is 3.8 × 10^{-8} cm, the molar mass is 127.9 g mol^{-1}, the Avogadro number is 6.023 × 10^{23} mol L^{-1} and gas constant R = 8.314 J mol^{-1} k^{-1}.

(b) Calculate the steric factor if the experimental value of k = 2.0×10^{-6} L mol^{-1} s^{-1}.

SOLUTION

(a) (i) $k = Zpf$

Since $Z_{AA} = 4\pi r_A^2 \left(\dfrac{8k_B T}{\pi \mu_A}\right)^{\frac{1}{2}} N_A,\quad f = e^{-E_a/RT}$

$$\therefore k_{2HI} = 4\pi r_A^2 N_A \sqrt{\left(\dfrac{8k_B T}{\pi \mu_A}\right)} \cdot e^{-E_a/RT}$$

$$r = \dfrac{d_A}{2} = \dfrac{3.8 \times 10^{-8}}{2} = 1.9 \times 10^{-8}\ cm$$

$$N_A = 6.023 \times 10^{23},\quad T = 600K,\quad \mu_{HI} = \tfrac{1}{2} M_{HI} = \dfrac{127.9}{2},$$

$$k_B = 1.38 \times 10^{-23}\ JK^{-1},\quad E_a = 184.4\ Jmol^{-1},\quad R = 8.314\ Jmol^{-1}K^{-1}$$

$$k = 4\pi \left(1.9 \times 10^{-8}\right)^2 \left(6.023 \times 10^{23}\right) \sqrt{\dfrac{6.624 \times 10^{-23} \times 600}{\left(\pi \times \dfrac{127.9}{2}\right)}} \cdot e^{-184.4/8.314 \times 600}$$

$$k = 4\pi \left(3.61 \times 10^{-16}\right)\left(6.023 \times 10^{23}\right) \sqrt{\dfrac{8 \times 1.38 \times 10^{-23} \times 600}{\left(\pi \times \dfrac{127.9}{2}\right)}} \cdot e^{-\frac{184.4}{4988.7}}$$

$$k = 2.73 \times 10^9 \sqrt{3.33 \times 10^{-22}} \cdot e^{-0.037}$$

$$k = 2.73 \times 10^9 \left(1.82 \times 10^{-11}\right) \times 9.64 \times 10^{-1}$$

$$k = 4.79 \times 10^{-2}\ Lmol^{-1}s^{-1}$$

(a) (ii) The fraction of molecules activated is given by:

$$f = e^{-Ea/RT}$$

$$f = e^{-184.4/(8.3145 \times 600)}$$

$$f = e^{-\frac{184.4}{4988.7}} = e^{-0.037}$$

$$f = 9.64 \times 10^{-1}$$

$$f = 96.4\%$$

(b) From collision theory $k = Z_{AA}e^{-E_a/RT}$ eq(i)

Experimentally $k_e = pZ_{AA}e^{-E_a/RT}$ eq(ii)

By substituting equation (i) into equation (ii) we have:

$$k_e = pk$$

$$p = k_e/k \quad (0 < p < 1)$$

Where p = steric factor, k = theoretical collision rate constant,

$$p = \frac{2.0 \times 10^{-6}}{4.97 \times 10^{-2}}$$

$$p = 4.02 \times 10^{-5}$$

EXERCISE 4

(1) What is the fraction of collision that has sufficient energy for a reaction if the activation energy is 50 kJ mol⁻¹ given that the temperature is … ?

(a) $25°C$ $\left[\text{Ans} : 1.7 \times 10^{-9} \right]$

(b) $500°C$ $\left[\text{Ans} : 4.2 \times 10^{-4} \right]$

(2) Food rots about 40 times more rapidly at 25° C than when it is stored at 4° C. Estimate the overall activation energy for the processes responsible for its decomposition.

$$\left[\text{Ans: } E_a = 122.68 \text{ KJ mol}^{-1} \right]$$

(3) What proportion of collisions between NO_2 molecules have enough energy to result in a reaction when the temperature is at ... ?
(a) 20° C
(b) 200° C
Given that activation energy E_a for this reaction below is 111 KJ mol^{-1},

$$2NO_{2(g)} \rightarrow 2NO_{(g)} + O_{2(g)}$$

$$\left[\text{Ans: (a) } f = 1.64 \times 10^{-20}, \text{ (b) } f = 5.55 \times 10^{-13} \right]$$

(4) Use collision theory to estimate the pre-exponential factor for the above reaction in (3) given that the experimental value of k is 2.0×10^9 L mol^{-1} s^{-1}.

$$[\text{Ans}: \text{(a) } A = 1.22 \times 10^{29} \text{ (b) } A = 3.6 \times 10^{21} \text{ ; Hint use k = Af}$$
where A = pre-exponential factor and $f = e^{-E_a/RT}$]

(5) The activation energy of a reaction is 65 KJ mol^{-1}. The rate constant at 250 K is 3.1×10^{-3} s^{-1}. At what temperature will the rate constant be 5.4×10^{-2} s^{-1}?

$$\left[\text{Ans}: T = 276\,K; \text{ use } T_2 = \left[\frac{1}{T_1} - \frac{R\ln(k_2/k_1)}{E_a} \right]^{-1} \right]$$

(6) The rate of a particular reaction quadruples when the temperature is increased from 25° C to 35° C. Calculate the activation energy for this reaction:

$$\left[\text{Ans}: E_a = 104.78 \text{ KJ mol}^{-1} \right]$$

(7) The activation energy of a reaction is 105 KJ mol^{-1}. If the rate constant is increased by four times when the temperature changes from 25° C to a new temperature, what is the final temperature?

$$\left[\text{Ans}: T_2 = 35\,°C \right]$$

5 Steady-State Approximation, Reaction Mechanism and Rate Law of Chain Reactions

5.1 STEADY-STATE APPROXIMATION

Steady-state approximation proposes that the concentration of all intermediates remains constant and small throughout the reaction. An intermediate is any species that does not appear in the overall reaction but has been introduced into the mechanism of that reaction. In other words, an intermediate is a species that is produced and consumed during a chemical reaction. The steady-state approximation is the most common approach to the analysis of reaction mechanisms. Hence, the principle assumes that

Rate of formation of intermediate = Rate of consumption of intermediate

If a concentration of intermediate species $= [I]$

Then, $\dfrac{d[I]}{dt} = -\dfrac{d[I]}{dt}$

5.2 REACTION MECHANISM

Most reactions will not proceed in a single step. As such, the step-by-step pathway by which a reaction occurs is referred to as the reaction mechanism. Every individual step in the mechanism is called an elementary reaction.

Elementary reactions are characterized by the following:

(a) They are either unimolecular (involves a single reactant species) or bimolecular (involves two reactant molecules), but termolecular (involves three reactant molecules colliding simultaneously) are very rare.

DOI: 10.1201/9781003274384-5

(b) The order of an elementary reaction corresponds with the stoichiometric coefficients in the balanced equation for that step, although
this is not always true for the overall rate law and overall balanced
equation in an experiment.

(c) The overall rate law of a reaction depends on a single elementary
step which is usually the slowest step in the reaction mechanism.

(d) They can also be reversible reactions having both forward and
reverse processes at equilibrium.

5.3   RATE LAW OF CHAIN REACTION

Chain reactions often lead to complex reactions with complicated rate laws.
The mechanism of a chain reaction is composed of initiation, propagation,
retardation and termination steps. The intermediate, also known as a chain
carrier, is generated in the initiation step. The chain carrier produced then
attacks other reactant molecules in the propagation step and each attack
gives rise to a new chain carrier. In the retardation step, a chain carrier
attacks a product molecule that was earlier formed between chain carriers
(free radical intermediates) to end a chain reaction.

5.4   ANALYSIS OF REACTION MECHANISMS
USING STEADY-STATE APPROXIMATION

(i) Examine the proposed mechanism for whether it satisfies the balanced equation of the overall reaction.

(ii) Write the expression for the reaction rate using the rate-determining (slow) step in the reaction mechanism.

(iii) Determine the steady-state concentration of all intermediates by
applying the steady-state approximation:

$$\frac{d[I]}{dt} \text{ produced} = -\left(\frac{d[I]}{dt}\right) \text{consumed}$$

(iv) Substitute the intermediate concentration into the rate expression
of the rate-determining step to eliminate the intermediate.

Note: If the slowest step in the reaction mechanism does not involve any
intermediate, then the rate law will not require a steady-state concentration
of intermediates.

It is important to also differentiate between an intermediate species and
catalyst in a reaction mechanism. An intermediate species is any species
that does not appear in the overall chemical reaction but is produced and
simultaneously consumed in the elementary steps of a reaction mechanism.

On the contrary, a catalyst is a species that appears unchanged in a reaction. This means it is a species that is present at the beginning of a reaction (first elementary step) and as a product at the end of a reaction (last elementary step).

Example 1: The reaction of ozone with nitrogen dioxide to produce oxygen and dinitrogen pentoxide is given as:

$$O_{3(g)} + 2NO_{2(g)} \rightarrow O_{2(g)} + N_2O_{5(g)}$$

The proposed mechanism is:

$$O_3 + NO_2 \xrightarrow{\ k_1\ } NO_3 + O_2 \quad (\text{Slow})$$

$$NO_3 + NO_2 \xrightarrow{\ k_2\ } N_2O_5 \quad (\text{Fast})$$

What is the predicted rate law by this mechanism?

SOLUTION

The rate determining step is the slow step with the rate expression

$$\text{Rate} = k_1 [O_3][NO_2]$$

Since the intermediate NO_3 is not present in the rate law, therefore, $\text{Rate} = k_1 [O_3][NO_2]$

Example 2: The iodide ion catalyzed decomposition of hydrogen peroxide H_2O_2 is believed to follow the mechanism.

$$H_2O_2 + I^- \xrightarrow{\ k_1\ } H_2O + IO^- \quad (\text{slow})$$

$$H_2O_2 + IO^- \xrightarrow{\ k_1\ } H_2O + O_2 + I^- \quad (\text{fast})$$

SOLUTION

The intermediate is IO^-.
   The rate-determining step has the rate expression:

$$\text{Rate} = k_1 [H_2O_2][I^-]$$

Since the intermediate IO^- is not present in the rate expression,

$$\text{Therefore, Rate} = k_1[H_2O_2][I^-]$$

Example 3: The following mechanism has been proposed for the decomposition of ozone in the atmosphere:

$$O_3 \underset{k_1'}{\overset{k_1}{\rightleftharpoons}} O_2 + O \quad (\text{fast})$$

$$O + O_3 \xrightarrow{k_2} O_2 + O_2 \quad (\text{slow})$$

Use the steady-state approximation, with O treated as the intermediate, to find the rate law of decomposition of O_3. Show that the rate is second-order in O_2.

SOLUTION

From the slow step, $O_3 + O \xrightarrow{k_2} O_2 + O_2$

$$\text{Rate} = k_2[O][O_3]$$

But [O] is an intermediate:

$$\frac{d[O]}{dt} = \text{Rate of formation of } [O] - \text{Rate of consumption of } [O]$$

From the fast equilibrium step,

$$\frac{d[O]}{dt} = k_1[O_3] - k_1'[O_2][O]$$

According to steady-state approximation:

$$\frac{d[O]}{dt} = k_1[O_3] - k_1'[O_2][O] = 0$$

$$\frac{k_1[O_3]}{k_1'[O_2]} = \frac{k_1'[O_2][O]}{k_1'[O_2]}$$

Substituting the intermediate concentration into the rate law:

$$\text{Rate} = k_2 \left(\frac{k_1[O_3]}{k_1'[O_2]} \right)[O_3]$$

$$\text{Rate} = \frac{k_1 k_2 [O_3][O_3]}{k_1'[O_2]} = \frac{k_1 k_2}{k_1'}[O_3]^2[O_2]^{-1}$$

$$\text{Rate} = \frac{k_1 k_2}{k_1'}[O_3]^2[O_2]^{-1}$$

Example 4: Urea, $(NH_2)_2CO$, can be prepared by heating ammonium cyanate, NH_4OCN.

$$NH_4OCN \rightarrow (NH_2)_2 CO$$

This reaction may occur by the following mechanism:

$$NH^+_4 + OCN^- \underset{k_{-1}}{\overset{k_1}{\rightleftharpoons}} NH_3 + HOCN \quad \text{(fast, equilibrium)}$$

$$NH_3 + HOCN \overset{k_2}{\longrightarrow} (NH_2)_2 CO \quad \text{(slow)}$$

What is the rate law predicted by this mechanism?

SOLUTION

From the slow step, $NH_3 + HOCN \overset{k_2}{\longrightarrow} (NH_2)_2 CO$

$$\text{Rate} = k_2 [NH_3][HOCN]$$

But $[NH_3]$ and $[HOCN]$ are both intermediates,
   From the fast step,

$$NH^+_4 + OCN^- \underset{k_{-1}}{\overset{k_1}{\rightleftharpoons}} NH_3 + HOCN$$

Using steady-state principle,

Rate of formation of intermediate – Rate of consumption

of intermediate $= 0$

$$k_1 [NH^+_4][OCN^-] - k_{-1}[NH_3][HOCN] = 0$$

$$\frac{k_1 [NH^+_4][OCN^-]}{k_{-1}} = \frac{\cancel{k_{-1}}[NH_3][HOCN]}{\cancel{k_{-1}}}$$

$$[NH_3][HOCN] = \frac{k_1}{k_{-1}}[NH^+_4][OCN^-]$$

Substituting intermediate concentrations into the rate law,

$$Rate = k_2\left(\frac{k_1}{k_{-1}}[NH^+_4][OCN^-]\right)$$

$$Rate = \frac{k_1k_2}{k_{-1}}[NH^+_4][OCN^-]$$

Example 5: The following mechanism has been proposed for the thermal decomposition of acetaldehyde (ethanal):

$$CH_3CHO \xrightarrow{k_a} \cdot CH_3 + CHO$$

$$\cdot CH_3 + CH_3CHO \xrightarrow{k_b} CH_4 + \cdot CH_2CHO$$

$$\cdot CH_2CHO \xrightarrow{k_c} CO + \cdot CH_3$$

$$\cdot CH_3 + \cdot CH_3 \xrightarrow{k_d} CH_3CH_3$$

Find the rate expression for the rate of formation of ethane and disappearance of ethanol.

SOLUTION

From elementary step (ii), $\cdot CH_3 + CH_3CHO \xrightarrow{k_b} CH_4 + \cdot CH_2CHO$

$$-\frac{d[CH_3CHO]}{dt} = +\frac{d[CH_4]}{dt} \qquad\qquad \text{eq (i)}$$

The rate of formation of methane is obtained from step (ii) above.

$$\frac{d[CH_4]}{dt} = k_b[CH_3CHO][\cdot CH_3] \qquad\qquad \text{eq (ii)}$$

Since $[\cdot CH_3]$ is an intermediate species then we eliminate it by steady-state approximation for all the intermediates.
$[\cdot CH_3]$ and $[\cdot CH_2CHO]$ are intermediates in the mechanism.

Net rate of $[\cdot CH_3]$ = Rate of

$[\cdot CH_3]$ formation – Rate of $[\cdot CH_3]$ consumption

$$\frac{d[\cdot CH_3]}{dt} = k_a [CH_3CHO] - k_b [\cdot CH_3][CH_3CHO] + k_c [\cdot CH_2CHO]$$

$$- 2\left(k_d \left[[\cdot CH_3]^2\right]\right)$$

$$\frac{d[\cdot CH_3]}{dt} = k_a [CH_3CHO] - k_b [\cdot CH_3][CH_3CHO]$$

$$+ k_c [\cdot CH_2CHO] - 2k_d [\cdot CH_3]^2 .$$

eq (iii)

Net rate of $[\cdot CH_2CHO]$ = Rate of $[\cdot CH_2CHO]$ formation – Rate of $[\cdot CH_2CHO]$ consumption

$$\frac{d[\cdot CH_2CHO]}{dt} = k_b [CH_3CHO][\cdot CH_3] + k_c [\cdot CH_2CHO] \quad \text{eq(iv)}$$

$$\frac{d[\cdot CH_3]}{dt} + \frac{d[\cdot CH_2CHO]}{dt} = 0 \quad \text{(Steady state approximation)}$$

$$k_a [CH_3CHO] - \cancel{k_b [\cdot CH_3][CH_3CHO]} + \cancel{k_c [\cdot CH_2CHO]}$$

$$- 2k_d [\cdot CH_3]^2 + \cancel{k_b [CH_3CHO][\cdot CH_3]} - \cancel{k_c [\cdot CH_2CHO]} = 0$$

$$k_a [CH_3CHO] - 2k_d [\cdot CH_3]^2 = 0$$

$$\frac{k_a [CH_3CHO]}{2k_d} = \frac{\cancel{2k_d} [\cdot CH_3]^2}{\cancel{2k_d}}$$

$$[\cdot CH_3]^2 = \frac{k_a}{2k_d}[CH_3CHO]$$

$$[\cdot CH_3]^{2\times\frac{1}{2}} = \left(\frac{k_a}{2k_d}[CH_3CHO]\right)^{\frac{1}{2}}$$

$$[\cdot CH_3] = \left(\frac{k_a}{2k_d}\right)^{\frac{1}{2}} ([CH_3CHO])^{\frac{1}{2}} \qquad\qquad eq\ (v)$$

Substituting $[\cdot CH_3]$ from equation (v) into equation (ii),

$$\frac{d[CH_4]}{dt} = k_b [CH_3CHO]\left(\frac{k_a}{2k_d}\right)^{\frac{1}{2}} ([CH_3CHO])^{\frac{1}{2}}$$

$$\frac{d[CH_4]}{dt} = k_b \left(\frac{k_a}{2k_d}\right)^{\frac{1}{2}} [CH_3CHO][CH_3CHO]^{\frac{1}{2}}$$

$$\frac{d[CH_4]}{dt} = k_b \left(\frac{k_a}{2k_d}\right)^{\frac{1}{2}} [CH_3CHO]^{\frac{3}{2}}$$

From equation (i), $-\dfrac{d[CH_3CHO]}{dt} = k_b \left(\dfrac{k_a}{2k_d}\right)^{\frac{1}{2}} [CH_3CHO]^{\frac{3}{2}}$

Therefore,

$$\frac{d[CH_3CHO]}{dt} = -k_b \left(\frac{k_a}{2k_d}\right)^{\frac{1}{2}} [CH_3CHO]^{\frac{3}{2}}$$

EXERCISE 5

(1) Thermal decomposition of nitryl chloride $2NO_2Cl_{(g)} \rightarrow 2NO_2 + Cl_{2(g)}$ has a proposed mechanism shown in the following equations:

$$NO_2Cl \xrightarrow{\ k_1\ } NO_2 + Cl \quad (slow)$$

$$NO_2Cl + Cl \xrightarrow{\ k_2\ } NO_2 + Cl_2 \quad (fast)$$

What is the rate law by this mechanism?

$$\left[Ans;\ Rate = k_1[NO_2Cl] \quad Hint: slow\ step\ is\ the\ rate\ determining\ step \right]$$

(2) The reaction $H_{2(g)} + I_{2(g)} \rightarrow 2HI_{(g)}$ may occur by the following mechanism:

$$I_2 \underset{k_{-1}}{\overset{k_1}{\rightleftharpoons}} 2I \quad (fast)$$

$$I + I + H_2 \xrightarrow{k_2} 2HI \quad (\text{slow})$$

What is the rate law predicted by the mechanism?

$$\left[\text{Ans}: \text{Rate} = k_2 k_1 / k_{-1} [H_2][I_2]\right]$$

(3) A possible mechanism for a gas-phase reaction is given below. What is the rate law predicted by the mechanism?

$$NO + Cl_2 \underset{k_{-1}}{\overset{k_1}{\rightleftharpoons}} NOCl_2 \quad (\text{fast})$$

$$NOCl_2 + NO \xrightarrow{k_2} 2NOCl \quad (\text{slow})$$

$$\left[\text{Ans}: \text{Rate} = k_2 k_1 / k_{-1} [NO]^2 [Cl_2]\right]$$

(4) Tert-butyl chloride reacts in basic solution according to the equation.

$$(CH_3)_3 CCl + OH^- \rightarrow (CH_3)_3 COH + Cl^-$$

The accepted mechanism for this reaction is

$$(CH_3)_3 CCl \xrightarrow{k_1} (CH_3)_3 C^+ + Cl^- \quad (\text{slow})$$

$$(CH_3)_3 C^+ + OH^- \xrightarrow{k_2} (CH_3)_3 COH \quad (\text{fast})$$

What should be the rate law for this reaction?

$$\left[\text{Ans}: \text{Rate} = k_1 \left[(CH_3)_3 CCl\right]\right]$$

(5) The following mechanism has been proposed for the decomposition of acetaldehyde:

$$CH_3CHO \xrightarrow{k_i} \cdot CH_3 + CHO$$

$$\cdot CH_3 + CH_3CHO \xrightarrow{k_p} CH_4 + CH_3CO \cdot$$

$$CH_3CHO \cdot \xrightarrow{k_p'} \cdot CH_3 + CO$$

$$\cdot CH_3 + \cdot CH_3 \xrightarrow{k_t} CH_3CH_3$$

Find the rate law for the formation of methane.

$$\left[\text{Ans; } d[CH_4]/dt = k_p \left(\frac{k_i}{2k_t} \right)^{1/2} [CH_3CHO]^{3/2} \quad \text{OR} \quad k_r[CH_3CHO]^{3/2} \right]$$

(6) The reaction mechanism below involves an intermediate A. Deduce the rate law for the reaction.

$$A_2 \underset{k_{-1}}{\overset{k_1}{\rightleftharpoons}} A + A \quad (\text{fast})$$

$$A + B \xrightarrow{k_2} P \quad (\text{slow})$$

$$\left[\begin{array}{l} \text{Ans : Rate} = k_2 \left(\frac{k_1}{k_{-1}} \right)^{1/2} [A_2]^{1/2} [B] \\ \text{Hint; use the slow step to get the rate law and steady-state} \\ \text{concentration of intermediate} \end{array} \right]$$

(7) Consider the mechanism for a reaction in aqueous solution below.

$$NH_2NO_2 + OH^- \underset{k_{-1}}{\overset{k_1}{\rightleftharpoons}} H_2O + NHNO^-_2 \quad (\text{fast})$$

$$NH_2NO_2^- \xrightarrow{k_2} N_2O + OH^- \quad (\text{slow})$$

Assuming $[H_2O]$ is unitary ($[H_2O] = 1$), find the rate law for the reaction.

$$\left[\text{Ans : Rate} = k_2 k_1 / k_{-1} [NH_2NO_2][OH^-] \right]$$

Appendix
Solutions to Exercises

EXERCISE 1

(1) $H_{2(g)} + 2NO_{(g)} \rightarrow N_2O_{(g)} + H_2O_{(g)}$

Rate of NO disappearance = Rate of N_2O appearance

$$-\frac{d[NO]}{2dt} = +\frac{d[N_2O]}{dt}$$

$$-\frac{1}{2}\left(\frac{d[NO]}{dt}\right) = +\left(\frac{d[N_2O]}{dt}\right)$$

$$-\frac{1}{2}\left(-0.01\,\text{mol dm}^3\,\text{s}^{-1}\right) = \frac{d[N_2O]}{dt}$$

$$\frac{d[N_2O]}{dt} = \frac{0.01}{2}\,\text{mol dm}^{-3}\,\text{s}^{-1} = 5.0 \times 10^{-3}\,\text{mol dm}^{-3}\,\text{s}^{-1}$$

(2) (i) $N_{2(g)} + 3H_{2(g)} \rightarrow 2NH_{3(g)}$

(ii) $CH_{4(g)} + 2O_{2(g)} \rightarrow 2H_2O_{(g)} + 2CO_{2(g)}$

(3) Rate $= k[A]^1[B]^2$

$$\frac{[\text{concentration}]}{\text{time}} = k[A][B]^2$$

$$\frac{\text{mol dm}^{-3}}{\text{min.}} = k[\text{mol dm}^{-3}][\text{mol dm}^{-3}]^2$$

$$\frac{1}{(\text{mol dm}^{-3})^2\,\text{min}} = \frac{K(\text{mol dm}^{-3})^2}{(\text{mol dm}^{-3})^2}$$

$$k = (\text{mol dm}^{-3})^2\,\text{min}^{-1}$$

(4) $N_{2(g)} + 3H_{2(g)} \rightarrow 2NH_{3(g)}$

$$Rate = k[N_2]^1[H_2]^2$$

(5) $2NOCl_{(g)} \xrightarrow{\Delta} 2NO_{(g)} + Cl_{2(g)}$

$$Rate = -\frac{d[NOCl]}{2dt} = +\frac{d[NO]}{2dt} = +\frac{d[Cl_2]}{dt}$$

(a) Between the second and fourth minute:

$$Rate = -\frac{d[NOCl]}{2dt} = -\frac{1}{2}\left(\frac{0.0055 - 0.0071}{4-2}\right) mol\,dm^{-3}\,min^{-1}$$

$$Rate = -\frac{1}{2}\left(\frac{0.0016}{2}\right) = 4.0 \times 10^{-4}\,mol\,dm^{-3}\,min^{-1}$$

(b) Over the entire reaction (from 0 min to 10 min) using:

$$Rate = -\frac{1}{2}\left(\frac{d[NOCl]}{dt}\right) = -\frac{1}{2}\left(\frac{0.0033 - 0.010}{10-0}\right) mol\,dm^{-3}\,min^{-1}$$

$$Rate = -\frac{1}{2}\left(\frac{0.0067}{10}\right) = 3.4 \times 10^{-4}\,mol\,dm^{-3}\,min^{-1}$$

(6) $2 \cdot CH_{3(g)} \rightarrow CH_3 - CH_3$

$$Rate = -\frac{d[\cdot CH_3]}{2dt} = +\frac{d[CH_3CH_3]}{dt}$$

(a) Using

$$Rate = -\frac{d[\cdot CH_3]}{2dt} = -\frac{1}{2}(-1.2)mol\,L^{-1}\,s^{-1}$$

$$Rate = \frac{1.2}{2} = 0.6\,mol\,L^{-1}\,s^{-1}$$

(b) Using $Rate = +\dfrac{d[CH_3CH_3]}{dt}$

$$\frac{d[CH_3CH_3]}{dt} = +0.6\,mol\,L^{-1}\,s^{-1}$$

(7) $N_{2(g)} + 3H_{2(g)} \rightarrow 2NH_{3(g)}$

$$Rate = -\frac{d[N_2]}{dt} = -\frac{d[H_2]}{3dt} = +\frac{d[NH_3]}{2dt}$$

Rate of disappearance of H_2 = Rate of formation of NH_3

$$-\frac{1}{3}\left(\frac{d[H_2]}{dt}\right) = +\frac{1}{2}\left(\frac{d[NH_3]}{dt}\right)$$

$$\frac{d[H_2]}{dt} = -\frac{3}{2}\left(\frac{d[NH_3]}{dt}\right) = -\frac{3}{2}(0.345)Ms^{-1}$$

$$\frac{d[H_2]}{dt} = -0.518\,Ms^{-1}$$

EXERCISE 2

(1) Reaction; $H_{2(g)} + Cl_{2(g)} \rightarrow 2HCl_{(g)}$

   (a) Rate $= k[H_2]^x [Cl_2]^y$

        comparing experiments 1 and 3

$$R_1 = k[H_2]_1^x [Cl_2]_1^y \qquad\qquad\qquad \text{eq (i)}$$

$$R_3 = k[H_2]_3^x [Cl_2]_3^y \qquad\qquad\qquad \text{eq (ii)}$$

$$3.5\times10^{-3} = k[0.15]^x [0.075]^y$$

$$7.0\times10^{-3} = k[0.15]^x [0.15]^y$$

        Dividing the two equations for Experiments 1 and 3:

$$\frac{7.0\times10^{-3}}{3.5\times10^{-3}} = \frac{\cancel{k}\cancel{[0.15]}^{\cancel{x}}[0.15]^y}{\cancel{k}\cancel{[0.15]}^{\cancel{x}}[0.075]^y}$$

$$\frac{7.0}{3.5} = \left(\frac{0.15}{0.075}\right)^y$$

$$2 = 2^y$$

$$\cancel{2} = \cancel{2}^y$$

$$\therefore y = 1 \qquad \left([Cl_2] \text{ is } 1^{st} \text{ order} \right)$$

By dividing Experiments 2 and 3:

$$\frac{14 \times \cancel{10}^{-3}}{7.0 \times \cancel{10}^{-3}} = \frac{\cancel{k}[0.30]^x \cancel{[0.15]^y}}{\cancel{k}[0.15]^x \cancel{[0.15]^y}}$$

$$\frac{14}{7.0} = \left(\frac{0.30}{0.15} \right)^x$$

$$2 = 2^x$$

$$\cancel{2} = \cancel{2}^x$$

$$\therefore x = 1 \qquad \left([Cl_2] \text{ is } 1^{st} \text{ order} \right)$$

Hence, overall order = second order.

$$\text{Rate} = k[H_2][Cl_2]$$

(c) From the above rate expression:

$$k = \frac{\text{Rate}}{[H_2][Cl_2]}$$

using Experiment 2, Rate $= 1.4 \times 10^{-2} \text{ mol dm}^{-3} \text{ s}^{-1}$,
$[H_2] = 0.30 \text{ mol dm}^{-3}$, $[Cl_2] = 0.15 \text{ mol dm}^{-3}$

$$k = \frac{1.4 \times 10^{-2} \cancel{\text{mol dm}}^{-3} \text{s}^{-1}}{(0.30)(0.15) \left(\text{mol dm}^{-3} \right)^{\cancel{2}}} = \frac{1.4 \times 10^{-2}}{0.045} \text{mol}^{-1} \text{dm}^3 \text{s}^{-1}$$

$$k = 3.11 \times 10^{-1} \text{mol}^{-1} \text{dm}^3 \text{s}^{-1}$$

(2) Reaction; $SO_2 + O_3 \rightarrow SO_3 + O_2$

(a) Rate $= k[SO_2]^a [O_3]^b$

Using Experiments 1 and 2:

$$\frac{0.118}{0.118} = \frac{\cancel{k}\cancel{[0.25]^a}[0.40]^b}{\cancel{k}\cancel{[0.25]^a}[0.20]^b}$$

$$1 = \left(\frac{0.40}{0.20}\right)^b$$

$$1 = 2^b$$

from indices $1 = 2^0$

$$\cancel{2}^0 = \cancel{2}^b$$

$$\therefore b = 0 \quad \left([O_3] = \text{zero order}\right)$$

Using Experiments 3 and 2:

$$\frac{1.062}{0.118} = \frac{\cancel{k}[0.75]^a \cancel{[0.20]^b}}{\cancel{k}[0.25]^a \cancel{[0.20]^b}} \Rightarrow \frac{1.062}{0.118} = \left(\frac{0.75}{0.25}\right)^a$$

$$9 = 3^a$$

From indices $9 = 3^2$:

$$\cancel{3}^2 = \cancel{3}^a$$

$$\therefore a = 2 \left([SO_2] = \text{second order}\right)$$

Overall order = second order.

(b) Rate $= k[SO_2]^2 [O_3]^0 = k[SO_2]^2$

(c) From the above rate expression:

$$k = \frac{\text{Rate}}{[SO_2]^2}$$

Using Experiment 1, Rate $= 0.118 \, mol \, L^{-1} \, s^{-1}$, $[SO_2] = 0.25 \, mol \, L^{-1}$

$$k = \frac{0.118 \, \cancel{mol \, L^{-1}} \, s^{-1}}{(0.25)^2 \, \cancel{(mol \, L^{-1})}^2} = \frac{0.118}{0.0625} mol^{-1} \, L \, s^{-1}$$

$$k = 1.89 \, mol^{-1} \, L \, s^{-1}$$

(3) Reaction; $3NO_{(g)} + Cl_{2(g)} \rightarrow 2NOCl_{(g)}$

$$\text{Rate} = k[NO]^m [Cl_2]^n$$

Using Experiments 1 and 2:

$$\frac{3.4 \times 10^{-4}}{8.5 \times 10^{-5}} = \frac{k[0.03]^m \, \cancel{[0.01]^n}}{k[0.015]^m \, \cancel{[0.01]^n}}$$

$$\frac{34 \times 10^{-5}}{8.5 \times 10^{-5}} = \left(\frac{0.03}{0.015}\right)^m \Rightarrow 4 = 2^m$$

$$\Rightarrow \cancel{2}^2 = \cancel{2}^m$$

$$\therefore m = 2 \qquad \left(NO = 2^{nd} \text{ order}\right)$$

Using Experiments 2 and 3:

$$\frac{3.4 \times 10^{-4}}{8.5 \times 10^{-5}} = \frac{\cancel{k[0.015]^m} \, [0.04]^n}{\cancel{k[0.015]^m} \, [0.01]^n}$$

$$\frac{34 \times 10^{-5}}{8.5 \times 10^{-5}} = \left(\frac{0.04}{0.01}\right)^n \Rightarrow 4 = 4^n$$

$$\Rightarrow 4^2 = 4^n$$

$$\therefore \; n = 2 \qquad \left(Cl_2 = 1^{st} \text{ order}\right)$$

$$\therefore \; \text{Rate} = k[NO]^2 [Cl_2]$$

(4) Reaction; $NH_{4(aq)}^+ + NO_{2(aq)}^- \rightarrow N_{2(g)} + 2H_2O_{(aq)}$

$$Rate = k\left[NH_4^+\right]^x \left[NO_2^-\right]^y$$

(a) Using Experiments 1 and 2:

$$\frac{2.50 \times 10^{-3}}{1.25 \times 10^{-3}} = \frac{k[0.50]^x [0.25]^y}{k[0.25]^x [0.25]^y}$$

$$\frac{2.50}{1.25} = \left(\frac{0.50}{0.25}\right)^x \Rightarrow 2 = 2^x$$

$$\Rightarrow 2^1 = 2^x$$

$$\therefore x = 1 \quad \left(NH_4^+ = 1^{st} \text{ order}\right)$$

Using Experiments 1 and 3:

$$\frac{6.25 \times 10^{-4}}{12.5 \times 10^{-4}} = \frac{k[0.25]^x [0.125]^y}{k[0.25]^x [0.25]^y}$$

$$\frac{6.25}{12.5} = \left(\frac{0.125}{0.25}\right)^y \Rightarrow 0.5 = 0.5^y$$

$$\Rightarrow 0.5^1 = 0.5^y$$

$$\therefore y = 1 \quad \left(NO_2^- = 1^{st} \text{ order}\right)$$

$$\therefore Rate = k\left[NH_4^+\right]\left[NO_2^-\right]$$

(b) From the above rate law:

$$k = \frac{Rate}{\left[NH_4^+\right]\left[NO_2^-\right]}$$

Using experiment 1, $Rate = 1.25 \times 10^{-3} \, Ms^{-1}$, $\left[NH_4^+\right] = 0.25 M$, $\left[NO_2^-\right] = 0.25 \, M$

$$k = \frac{1.25 \times 10^{-3}\,\mathrm{M\,s^{-1}}}{(0.25)(0.25)\mathrm{M^2}} = \frac{1.25 \times 10^{-3}}{0.0625}\,\mathrm{M^{-1}\,s^{-1}}$$

$$k = 0.02 = 2.0 \times 10^{-2}\,\mathrm{M^{-1}\,s^{-1}}$$

(5) Reaction: $2N_2O_{5(g)} \rightarrow 4NO_{2(g)} + O_{2(g)}$

$$\text{Rate} = k\left[N_2O_5\right]^x$$

$$k = \frac{\text{Rate}}{\left[N_2O_5\right]^x}$$

Using Experiments 1 and 2,

$$\frac{45.0}{22.5} = \frac{\cancel{k}\left[2.56 \times \cancel{10^2}\right]^x}{\cancel{k}\left[1.28 \times \cancel{10^2}\right]^x}$$

$$\frac{45.0}{22.5} = \left(\frac{2.56}{1.28}\right)^x \Rightarrow 2 = 2^x$$

$$\Rightarrow \cancel{2}^1 = \cancel{2}^x$$

$$\therefore x = 1$$

$$\therefore \text{Rate} = k\left[N_2O_5\right]$$

Using Experiment 1, Rate $= 22.5\,\mathrm{M\,s^{-1}}$ and $\left[N_2O_5\right] = 1.28 \times 10^2\,\mathrm{M}$

$$k = \frac{\text{Rate}}{\left[N_2O_5\right]} = \frac{22.5\,\cancel{\mathrm{M}}\,\mathrm{s^{-1}}}{1.28 \times 10^2\,\cancel{\mathrm{M}}}$$

$$k = 0.176\,\mathrm{s^{-1}}$$

$$k = 1.76 \times 10^{-1}\,\mathrm{s^{-1}}$$

(6) Reaction: $H_{2(g)} + 2ICl_{(g)} \rightarrow I_{2(g)} + 2HCl_{(g)}$

$$\text{Rate} = k\left(P_{H_2}\right)^m \left(P_{ICl}\right)^n$$

(a) Using Experiments 1 and 2,

$$\frac{1.34}{0.331} = \frac{k(250)^m (325)^n}{k(250)^m (81)^n}$$

$$\frac{1.34}{0.331} = \left(\frac{325}{81}\right)^n \Rightarrow 4.05 = (4.01)^n$$

Take the logarithm of both sides:

$$\log 4.05 = \log(4.01)^n$$

$$\frac{\log 4.05}{\log 4.01} = \frac{n \cancel{\log 4.01}}{\cancel{\log 4.01}} \Rightarrow n = \frac{\log 4.05}{\log 4.01} = 1.0$$

$$\therefore \text{ICl} = 1^{st} \text{ order}$$

Using Experiments 1 and 3,

$$\frac{1.34}{0.266} = \frac{k(250)^m \cancel{(325)^n}}{k(50)^m \cancel{(325)^n}}$$

$$\frac{1.34}{0.266} = \left(\frac{250}{50}\right)^m \Rightarrow 5.04 = (5.0)^m$$

$$\log 5.04 = \log(5.0)^m$$

$$\frac{\log 5.04}{\log 5.0} = \frac{m \cancel{\log 5.0}}{\cancel{\log 5.0}} \Rightarrow m = \frac{\log 5.04}{\log 5.0} = 1.0$$

$$\therefore H_2 = 1^{st} \text{ order}$$

$$\text{Overall order} = 2^{nd} \text{ order}$$

$$\text{Rate} = k(P_{H_2})(P_{ICl})$$

(b) From the rate law above:

$$k = \frac{\text{Rate}}{(P_{H_2})(P_{ICl})}$$

Using Experiment 1, Rate $= 1.34\,\text{Torr}\,\text{s}^{-1}$, $P_{H_2} = 250\,\text{Torr}$, $P_{ICl} = 325\,\text{Torr}$

$$k = \frac{1.34\,\cancel{\text{Torr}}\,\text{s}^{-1}}{(250)(325)(\text{Torr})^{\cancel{2}}} = \frac{1.34}{81250}\,\text{Torr}^{-1}\,\text{s}^{-1}$$

$$k = 0.0000165\,\text{Torr}^{-1}\,\text{s}^{-1}$$

$$k = 1.65 \times 10^{-5}\,\text{Torr}^{-1}\,\text{s}^{-1}$$

EXERCISE 3

(1) $k = 0.636\,\text{s}^{-1}$

$$1^{\text{st}}\ \text{order half-life} = t_{\frac{1}{2}} = \frac{\ln 2}{0.366} = 1.09\,\text{s}$$

(2) Assuming $[A]_0 = 100\% = 1.0$

$$[A]_f = 100\% - 53\% = 47\% = 0.47$$

$$t = 30\,\text{min. } k = ?$$

Integrated rate law for first-order reactions:

$$[A]_f = [A]_0\,e^{-kt}$$

$$0.47 = 1.0e^{-k(30)}$$

$$0.47 = e^{-30k}$$

Natural logarithm of both sides:

$$\ln 0.47 = \ln e^{-30k}$$

$$\ln 0.47 = -30k\,\ln e$$

Since $\ln e = 1$:

$$\frac{\ln 0.47}{-30} = \frac{\cancel{-30}k}{\cancel{-30}}$$

$$k = \frac{-0.755}{-30} = 0.025\,\text{min}^{-1}$$

$$k = 2.52 \times 10^{-2}\,\text{min}^{-1}$$

(3) $[A]_0 = 0.397\,\text{M}$

$$[A]_f = ? \quad t = 100\,\text{s}$$

$$t_{1/2} = 19\,\text{s} \quad k = ?$$

Using half-life for first order $t_{1/2} = \dfrac{\ln 2}{k}$

$$k = \frac{\ln 2}{t_{1/2}} = \frac{\ln 2}{19} = 0.036\,\text{s}^{-1}$$

Integrated rate law for first order:

$$[A]_f = [A]_0\,e^{-kt}$$

$$[A]_f = 0.397 \times e^{-0.036 \times 100}$$

$$[A]_f = 0.397 \times 0.0273$$

$$[A]_f = 0.010\,\text{M}$$

$$[A]_f = 1.0 \times 10^{-2}\,\text{M}$$

(4) $2NO_{2(g)} \xrightarrow{\Delta} 2NO_{(g)} + O_{2(g)}$

$$\text{Rate} = k[NO_2]^2$$

Integrated rate law of second-order reaction:

$$\frac{1}{[A]_f} = kt + \frac{1}{[A]_0} \quad \text{or} \quad [A]_f = \frac{[A]_0}{[A]_0\,kt + 1}$$

Since $[A]_f = 0.56\,\text{M}$, $k = 7.8 \times 10^{-2}\,\text{M}^{-1}\,\text{s}^{-1}$, $t = 3.2\,\text{min} = 3.2 \times 60 = 192\,\text{s}$

$$[A]_f = \frac{0.56}{(0.56 \times 0.078 \times 192) + 1} = \frac{0.56}{8.387 + 1} = \frac{0.56}{9.387} M$$

$$[A]_f = 0.0597$$

$$[A]_f = 0.06\,M = 6.0 \times 10^{-2}\,M$$

(5) $t_{\frac{1}{2}} = 48\,s$ $[A]_0 = ?$ $k = 5.19 \times 10^{-2}\,M^{-1}\,s^{-1}$

$$2^{nd}\ \text{order half-life,}\ t_{\frac{1}{2}} = \frac{1}{[A]_0\,k}$$

$$\Rightarrow [A]_0 = \frac{1}{kt_{\frac{1}{2}}} = \frac{1}{0.519 \times 48}$$

$$[A]_0 = \frac{1}{2.49} = 0.40\,M$$

(6) $k = 2.5 \times 10^{-2}\,M^{-1}\,s^{-1}$, $t = ?$ $[A]_f = 100\% - 63\% = 37\% = 0.37$

$$[A]_0 = 100\% = 1.0$$

Integrated rate law second-order reactions:

$$[A]_f = \frac{[A]_0}{[A]_0\,kt + 1}$$

$$\Rightarrow t = \frac{[A]_0 - [A]_f}{[A]_0[A]_f\,k} = \frac{1.0 - 0.37}{(1.0)(0.37)(0.0025)}$$

$$t = \frac{0.63}{0.000925} = 681.1\,s$$

$$t = 681\,s$$

(7) $k = 0.47\,M^{-1}\,s^{-1}$, $[A]_0 = 0.53\,M$, $[A]_f = 0.13\,M$, $t = ?$

Integrated rate law for second order:

$$[A]_f = \frac{[A]_0}{[A]_0\,kt + 1}$$

$$\Rightarrow t = \frac{[A]_0}{[A]_0 [A]_f k} = \frac{0.53 - 0.13}{(0.53)(0.13)(0.47)}$$

$$t = \frac{0.4}{0.03241} = 12.34\,s$$

EXERCISE 4

(1) $f = e^{-E_a/RT}$

(a) When $T = 25°C = 298\,K$, $E_a = 50,000\,J\,mol^{-1}$, $R = 8.314\,J\,mol^{-1}\,K^{-1}$

$$f = e^{-50,000/8.314 \times 298}$$

$$f = e^{-50,000/2477.57} = e^{-20.18}$$

$$f = 1.7 \times 10^{-9}$$

(b) When $T = 500°C = 773\,K$, $E_a = 50,000\,J\,mol^{-1}$, $R = 8.314\,J\,mol^{-1}K^{-1}$

$$f = e^{-50,000/8.314 \times 773}$$

$$f = e^{-50,000/6426.72} = 4.18 \times 10^{-4}$$

$$f = 4.2 \times 10^{-4}$$

(2) Let the initial rate at $4°$ C be k.

Therefore the rate constant at $25°$ C will be 40k.

$$T_1 = 4°C = 277\,K, \ T_2 = 25°C = 298\,K$$

$$k_1 = k, \ \text{and} \ k_2 = 40k, \ R = 8.3145\,J\,mol^{-1}\,K^{-1}$$

Using the equation:

$$E_a = \frac{-R\ln(k_2/k_1)}{\left(\dfrac{1}{T_2} - \dfrac{1}{T_1}\right)}$$

$$E_a = \frac{-8,3145\ln(40k/k)}{\left(\dfrac{1}{298} - \dfrac{1}{277}\right)} = \frac{-8.3145\ln 40}{0.00336 - 0.0036}$$

$$E_a = \frac{-30.67}{-0.00025} = 122680 \, J \, mol^{-1}$$

$$E_a = 122.68 \times 10^3 \, J \, mol^{-1} = 122.68 \, KJ \, mol^{-1}$$

(3) $f = e^{-E_a/RT}$

(a) When $T = 20°C = 293k$, $E_a = 111 \, KJ \, mol^{-1}$, $R = 8.314 \, J \, mol^{-1} \, s^{-1}$

$$f = e^{-111000/8.314 \times 293}$$

$$f = e^{-111000/2436.0} = e^{-45.56}$$

$$f = 1.64 \times 10^{-20}$$

(b) When $T = 20°C = 473k$, $E_a = 111 \, kJ \, mol^{-1}$, $R = 8.314 \, J \, mol^{-1} \, k^{-1}$

$$f = e^{-111000/8.314 \times 293}$$

$$f = e^{-111000/3932.52} = e^{-28.22}$$

$$f = 5.55 \times 10^{-13}$$

(4) Using $k = Af$, $\quad k = 2.0 \times 10^9 \, L \, mol^{-1} \, s^{-1}$

(a) When $f = 1.64 \times 10^{-20}$:

$$A = k/_f = \frac{2.0 \times 10^9}{1.64 \times 10^{-20}}$$

$$A = 1.22 \times 10^{29}$$

(b) When $f = 5.55 \times 10^{-13}$:

$$A = \frac{k}{f} = \frac{2.0 \times 10^9}{5.55 \times 10^{-13}}$$

$$A = 0.36 \times 10^{22}$$

$$A = 3.6 \times 10^{21}$$

(5) $T_1 = 250\,K$ $k_1 = 0.0031\,s^{-1}$ $E_a = 65\,KJ\,mol^{-1} = 65,000\,J\,mol^{-1}$, $T_2 = ?$,

$$k_2 = 0.054\,s^{-1} \quad R = 8.314\,J\,mol^{-1}\,k^{-1}$$

Using:

$$T_2 = \left[\frac{1}{T_1} - \frac{R\ln(k_2/k_1)}{E_a} \right]^{-1}$$

$$T_2 = \left[\frac{1}{250} - \frac{8.314\ln(0.054/0.0031)}{65,000} \right]^{-1}$$

$$T_2 = \left[0.004 - (+0.000365) \right]^{-1}$$

$$T_2 = (0.00363)^{-1}$$

$$T_2 = 275.48\,K \cong 276\,K$$

(6) $T_1 = 25°C = 298\,K$, $k_1 = k$, $T_2 = 35°C = 308\,K$, $k_2 = 4k$, $E_a = ?$,

$R = 8.314\,J\,mol^{-1}\,K^{-1}$

Using:

$$E_a = \frac{-R\ln(k_2/k_1)}{\left(\dfrac{1}{T_2} - \dfrac{1}{T_1} \right)}$$

$$E_a = \frac{-8.314\ln(4\cancel{k}/\cancel{k})}{\left(\dfrac{1}{308} - \dfrac{1}{298} \right)} = \frac{-8.314\ln 4}{(0.00325 - 0.00336)}$$

$$E_a = \frac{-11.526}{-0.00011} = 104781.8\,J\,mol^{-1}$$

$$E_a = 104.78\,kJ\,mol^{-1}$$

(7) $E_a = 105\,kJ\,mol^{-1}$, $T_1 = 25°C = 298\,K$, $k_1 = k$, $T_2 = ?$, $k_2 = 4k$,
$R = 8.314\,kJ\,mol^{-1}\,s^{-1}$

$$T_2 = \left[\frac{1}{T_1} - \frac{R\ln(k_2 / k_1)}{E_a}\right]^{-1}$$

$$T_2 = \left[\frac{1}{298} - \frac{8.314\ln(4\cancel{k} / \cancel{k})}{105,000}\right]^{-1}$$

$$T_2 = [0.00336 - 0.000109]^{-1}$$

$$T_2 = [0.00325]^{-1}$$

$$T_2 = 307.7\,K \cong 308\,K$$

$$T_2 \text{ in } °C = 308 - 273 = 35°C$$

EXERCISE 5

(1) Overall reaction: $2NO_2Cl_{(g)} \rightarrow 2NO_{2(g)} + Cl_{2(g)}$
   Mechanism:

   (i) $NO_2Cl \xrightarrow{\ k_1\ } NO_2 + Cl$ (slow)

   (ii) $NO_2Cl + Cl \xrightarrow{\ k_2\ } NO_2 + Cl_2$ (fast)

   The slowest step is (i) above.
   Therefore, Rate $= k_1[NO_2Cl]$.

(2) Overall reaction: $H_{2(g)} + I_{2(g)} \rightarrow 2HI_{(g)}$
   Mechanism:

   (i) $I_2 \underset{k_{-1}}{\overset{k_1}{\rightleftharpoons}} 2I$ (fast)

   (ii) $I + I + H_2 \quad \overset{k_2}{\rightarrow} \quad 2HI$ (slow)

   The rate-determining step is step (ii) above.

   Therefore, Rate $= k_2[I][I][H_2] = k_2[I]^2[H_2]$

Since I is an intermediate and is not in the overall reaction, by applying steady-state approximation $\dfrac{d[I]}{dt} = 0$,

$$\frac{d[I]}{dt} = 2k_1[I_2] - 2k_{-1}[I]^2 = 0$$

$$\frac{2k_1[I_2]}{2k_{-1}} = \frac{\cancel{2k_{-1}}[I]^2}{\cancel{2k_{-1}}}$$

$$[I]^2 = \frac{2k_1[I_2]}{2k_{-1}} = \frac{k_1[I_2]}{k_{-1}}$$

Substituting $\dfrac{k_1[I_2]}{k_{-1}}$ for $[I]^2$ in the rate expression,

$$\text{Rate} = k_2\left(\frac{k_1[I_2]}{k_{-1}}\right)[H_2]$$

$$\text{Rate} = \frac{k_2 k_1}{k_{-1}}[I_2][H_2]$$

(3) Overall reaction: $2NO + Cl_2 \rightarrow 2NOCl$
   Mechanism:

(i) $NO + Cl_2 \underset{k_{-1}}{\overset{k_1}{\rightleftarrows}} NOCl_2$ (fast)

(ii) $NOCl_2 + NO \rightarrow 2NOCl$ (slow)

$$\text{Rate} = k_2[NOCl_2][NO]$$

Since $[NOCl_2]$ is an intermediate,

$$\frac{d[NOCl_2]}{dt} = 0 \quad (\text{steady state approximation})$$

$$\frac{d[NOCl_2]}{dt} = k_1[NO][Cl_2] - k_{-1}[NOCl_2] = 0$$

$$\frac{k_1[NO][Cl_2]}{k_{-1}} = \frac{\cancel{k_{-1}}[NOCl_2]}{\cancel{k_{-1}}}$$

$$[NOCl_2] = \frac{k_1[NO][Cl_2]}{k_{-1}}$$

Substituting $\dfrac{k_1}{k_{-1}}[NO][Cl_2]$ for [$NOCl_2$] into rate law,

$$Rate = k_2\left(\frac{k_1}{k_{-1}}\right)[NO][Cl_2][NO]$$

$$Rate = \frac{k_1 k_2}{k_{-1}} k_1 [NO]^2 [Cl_2]$$

(4) Overall reaction: $(CH_3)_3 CCl + OH^- \rightarrow (CH_3)_3 COH + Cl^-$

Mechanism:

(i) $(CH_3)_3 CCl \xrightarrow{k_1} (CH_3)_3 C^+ + Cl^-$ (slow)

(ii) $(CH_3)_3 C^+ + OH^- \xrightarrow{k_2} (CH_3)_3 COH$ (fast)

The slowest step is step (i).

Therefore, $Rate = k_1\left[(CH_3)_3 CCl\right]$

(5) Overall reaction: $CH_3CHO_{(g)} \rightarrow CH_{4(g)} + CO_{(g)}$

Mechanism:

(i) $CH_3CHO \xrightarrow{k_i} \cdot CH_3 + CHO$

(ii) $\cdot CH_3 + CH_3CHO \xrightarrow{k_p} CH_4 + CH_3\dot{C}O$

(iii) $CH_3\dot{C}O \xrightarrow{k'_p} \cdot CH_3 + CO$

(iv) $\cdot CH_3 + \cdot CH_3 \xrightarrow{k_t} CH_3CH_3$

$$\frac{d[CH_4]}{dt} = k_p[CH_3CHO][\cdot CH_3]$$

since [$\cdot CH_3$] is an intermediate.

The intermediate species in the mechanism are $\left[CH_3\dot{C}O\right]$ and $[\cdot CH_3]$.

$$\frac{d[\cdot CH_3]}{dt} = k_i[CH_3CHO] - k_p[CH_3CHO][\cdot CH_3] + k'_p[CH_3\dot{C}O] - 2k_t[\cdot CH_3]^2$$

$$\frac{d\left[CH_3C\dot{O}\right]}{dt} = k_p\left[CH_3CHO\right]\left[\cdot CH_3\right] + k'_p\left[CH_3C\dot{O}\right]$$

$$\frac{d\left[\cdot CH_3\right]}{dt} + \frac{d\left[CH_3C\dot{O}\right]}{dt} = 0 \quad \text{(Steady state approximation)}$$

$$k_i\left[CH_3CHO\right] - \cancel{k_p\left[CH_3CHO\right]\left[\cdot CH_3\right]} + \cancel{k'_p\left[CH_3C\dot{O}\right]}$$

$$- 2k_t\left[\cdot CH_3\right]^2 + k_p\left[CH_3CHO\right]\left[\cdot CH_3\right] - \cancel{k'_p\left[CH_3C\dot{O}\right]} = 0$$

$$k_i\left[CH_3CHO\right] - 2k_t\left[\cdot CH_3\right]^2 = 0$$

$$k_i\left[CH_3CHO\right] = 2k_t\left[\cdot CH_3\right]^2$$

$$\left[\cdot CH_3\right] = \left(\frac{k_i}{2k_t}\right)^{\frac{1}{2}} \left[CH_3CHO\right]^{\frac{1}{2}}$$

$$\frac{d\left[CH_4\right]}{dt} = k_p\left[CH_3CHO\right]\left(\frac{k_i}{2k_t}\right)^{\frac{1}{2}} \left[CH_3CHO\right]^{\frac{1}{2}}$$

$$\frac{d\left[CH_4\right]}{dt} = k_p\left(\frac{k_i}{2k_t}\right)^{\frac{1}{2}} \left[CH_3CHO\right]^{\frac{3}{2}}$$

if we let $k_p\left(\dfrac{k_i}{2k_t}\right)^{\frac{1}{2}} = k_r$

Then, $\dfrac{d\left[CH_4\right]}{dt} = k_r\left[CH_3CHO\right]^{\frac{3}{2}}$

(6) Mechanism:

$$A_2 \underset{k_{-1}}{\overset{k_1}{\rightleftarrows}} A + A \quad \text{(fast)}$$

$$A + B \xrightarrow{k_2} P \quad \text{(slow)}$$

Since [A] is an intermediate.

The rate law of the slow step is Rate $= k_2[A][B]$.

$$\frac{d[A]}{dt} = 2k_1[A_2] - 2k_{-1}[A]^2 = 0 \quad (\text{steady-state})$$

$$\frac{\cancel{2k_1}[A_2]}{\cancel{2k_{-1}}} = \frac{\cancel{2k_{-1}}[A]^2}{\cancel{2k_{-1}}}$$

$$[A]^2 = \frac{k_1}{k_{-1}}[A_2]$$

$$[A] = \left(\frac{k_1}{k_{-1}}\right)^{\frac{1}{2}}[A_2]^{\frac{1}{2}}$$

Substituting for [A] into the rate law.
Then,

$$\text{Rate} = k_2\left(\frac{k_1}{k_{-1}}\right)^{\frac{1}{2}}[A_2]^{\frac{1}{2}}[B]$$

$$\text{Rate} = k_2\left(\frac{k_1}{k_{-1}}\right)^{\frac{1}{2}}[A_2]^{\frac{1}{2}}[B]$$

(7) Mechanism:

(i) $NH_2NO_2 + OH^- \underset{k_{-1}}{\overset{k_1}{\rightleftharpoons}} H_2O + NHNO_2^- \quad (\text{fast})$

(ii) $NHNO_2^- \overset{k_2}{\longrightarrow} N_2O + OH^- \quad (\text{slow})$

The slow step has the rate law as:

$$\text{Rate} = k_2\left[NHNO_2^-\right]$$

since $\left[NHNO_2^-\right]$ is an intermediate.

$$\frac{d\left[NHNO_2^-\right]}{dt} = k_1[NH_2NO_2]\left[OH^-\right] - k_{-1}[H_2O]\left[NHNO_2^-\right] = 0$$

$$\frac{k_1[NH_2NO_2][OH^-]}{k_{-1}[H_2O]} = \frac{\cancel{k_{-1}}\cancel{[H_2O]}[NHNO_2^-]}{\cancel{k_{-1}}\cancel{[H_2O]}}$$

$$[NHNO_2^-] = \frac{k_1[NH_2NO_2][OH^-]}{k_{-1}[H_2O]}$$

Since $[H_2O] = 1$.

$$[NHNO_2^-] = \frac{k_1}{k_{-1}}[NH_2NO_2][OH^-]$$

$$[NHNO_2^-] = k_2\left(\frac{k_1}{k_{-1}}[NH_2NO_2][OH^-]\right)$$

Therefore,

$$\text{Rate} = \frac{k_2 k_1}{k_{-1}}[NH_2NO_2][OH^-]$$

Bibliography

Chang P. Chieh; Hana El-Samad; John D. Bookstaver and Dan Reid; A PowerPoint presentation on chemical kinetics. Retrieved on 18th May 2020. https://users.cs.duke.edu/~reif/courses/molcomplectures/Kinetics/KineticsOverview/KineticsOverview.ppt

Dorrell D. Ebbing and Stephen D. Gammon; *General Chemistry*; 9th Edition; Houghton Mifflin Company, New York (2009); pp. 219–237.

Jasperse kinetics practice problems. Retrieved on 18th March 2019. http://web.mnstate.edu/jasperse/Chem210/Extra%20Practice%20Sets%20Chem%20210/Test1%20ch15%20Kinetics%20%20Practice%20Problems.pdf

Mandes quiz on AP chemistry titled kinetics practice problems and solutions. Retrieved on 18th March 2021. http://www.delandhs.org/_cache/files/c/1/c10cd6c6-c726-4b39-87ad-83a9aed8427c/11280D23BF7E8BE93781F873F16905F8.kinetics-ws-3.pdf

Peter Atkins and Julio de Paula; *Atkins Physical Chemistry*; 7th Edition; Oxford University Press, New York (2002); pp. 862–919.

Peter Atkins and Julio de Paula; *Elements of Physical Chemistry*; 5th Edition; Oxford University Press, Britain (2009); pp. 219–237.

Walter J. Moore; *Physical Chemistry*; 4th Edition, Longmans Green and Co. Ltd, Britain (1963); pp. 253–312.

Index